A PLANET OF VIRUSES

CARL ZIMMER

A Planet of VIRUSES

SECOND EDITION

The University of Chicago Press | Chicago and London

CARL ZIMMER is a columnist for the *New York Times*, writes for *National Geographic* and other magazines, and is the author of thirteen books, including *Parasite Rex*, *Soul Made Flesh*, and *Microcosm*. He is a lecturer at Yale University, where he teaches writing about science and the environment.

The University of Chicago Press, Chicago 60637
The University of Chicago Press, Ltd., London

First edition published 2011. Second edition 2015.
Printed in the United States of America

24 23 22 21 20 19 18 17 16 15   1 2 3 4 5

ISBN-13: 978-0-226-29420-9 (paper)
ISBN-13: 978-0-226-32026-7 (e-book)
DOI: 10.7208/chicago/9780226320267.001.0001

The essays in this book were written for the World of Viruses project, funded by the National Center for Research Resources at the National Institutes of Health through the Science Education Partnership Award (SEPA) Grant No. R25 RR024267 (2007–2012). Its content is solely the responsibility of the authors and does not necessarily represent the official views of NCRR or NIH. Visit http://www.worldofviruses.unl.edu for more information and free educational materials about viruses. World of Viruses is a project of the University of Nebraska–Lincoln.

Library of Congress Cataloging-in-Publication Data

Zimmer, Carl, 1966- author.
A planet of viruses / Carl Zimmer. — Second edition.
pages ; cm
Includes bibliographical references and index.
ISBN 978-0-226-29420-9 (pbk. : alk. paper) — ISBN 978-0-226-32026-7 (e-book) 1. Viruses. I. Title.
QR360.Z65 2015
616.9´101—dc23

2015011313

♾ This paper meets the requirements of ANSI/NISO Z39.48-1992 (Permanence of Paper).

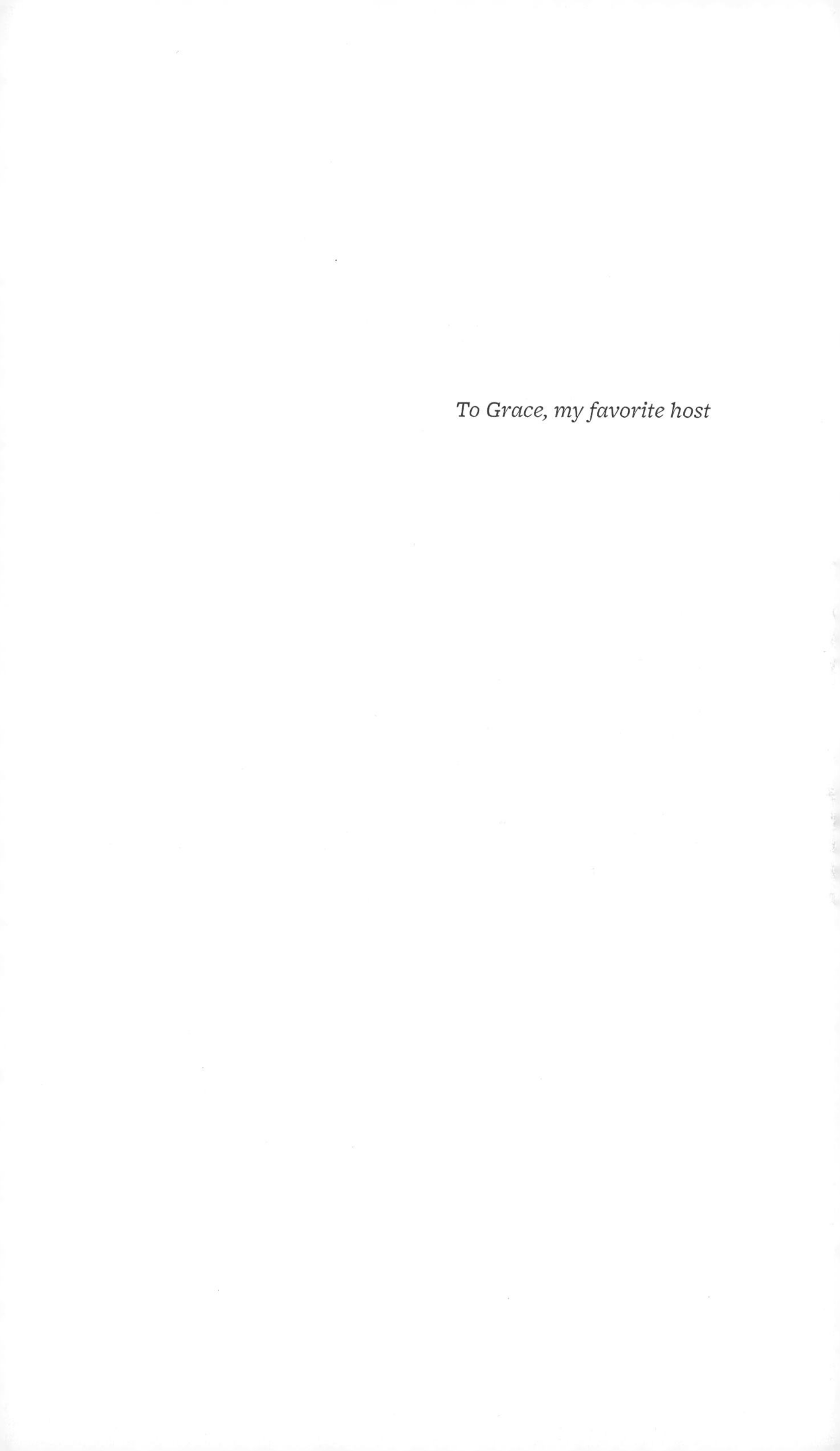

To Grace, my favorite host

Contents

Foreword

Viruses wreak chaos on human welfare, affecting the lives of almost a billion people. They have also played major roles in the remarkable biological advances of the past century. The smallpox virus was humanity's greatest killer, and yet it is now one of the few diseases to have been eradicated from the globe. New viruses, such as HIV, continue to pose new threats and challenges.

Viruses are unseen but dynamic players in the ecology of Earth. They move DNA between species, provide new genetic material for evolution, and regulate vast populations of organisms. Every species, from tiny microbes to large mammals, is influenced by the actions of viruses. Viruses extend their impact beyond species to affect climate, soil, the oceans, and fresh water. When you consider how every animal, plant, and microbe has been shaped through the course of evolution, one has to consider the influential role played by the tiny and powerful viruses that share this planet.

After the first edition of *A Planet of Viruses* was published in 2011, viruses continued to surprise us all. The Ebola virus, once limited to small flare-ups in remote parts of Africa, exploded into a massive outbreak in cities like Freetown and Conakry, and, for the first time, spread to other continents. New viruses, like MERS, leapt from animals to humans. But scientists also discovered new ways to harness the amazing diversity of viruses for our own benefit. Carl Zimmer has drawn on all these developments to produce this second edition of *A Planet of Viruses*.

Zimmer originally wrote these essays for the World of Viruses project as part of a Science Education Partnership Award (SEPA) from the National Center for Research Resources (NCRR) at the National Institutes of Health (NIH). World of Viruses was created to help people understand more about viruses and virology research through radio documentaries, graphic stories, teacher professional development, mobile phone and iPad applications, and other materials. For more information about World of Viruses, visit http://worldofviruses.unl.edu.

JUDY DIAMOND, PhD

Professor and Curator, University of Nebraska State Museum
Director of the World of Viruses Project

CHARLES WOOD, PhD

Lewis L. Lehr University Professor of Biological Sciences and Biochemistry
Director of the Nebraska Center for Virology

INTRODUCTION

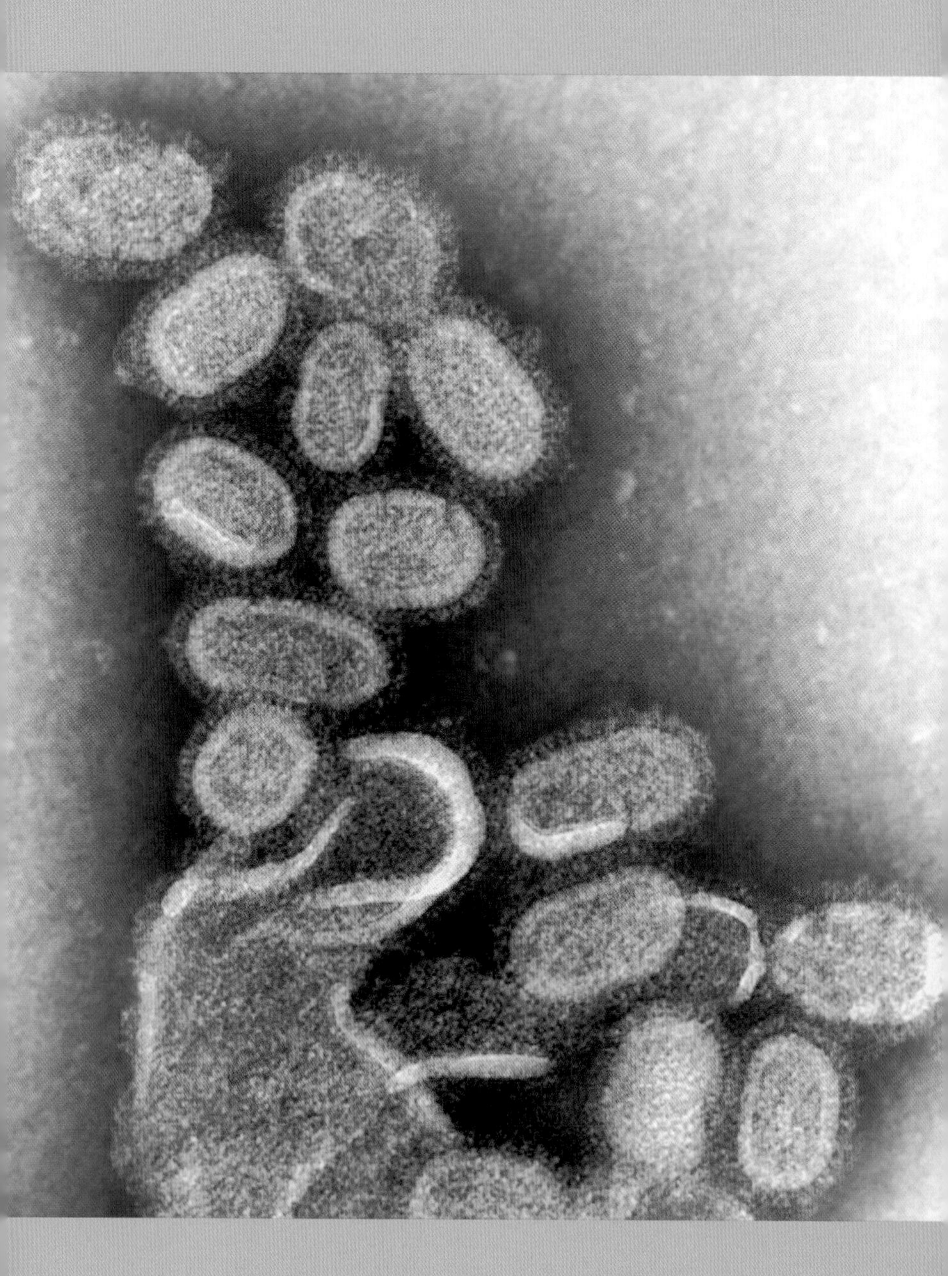

Tobacco mosaic viruses, which cause plant diseases worldwide

"A Contagious Living Fluid"

Tobacco Mosaic Virus and the Discovery of the Virosphere

Fifty miles southeast of the Mexican city of Chihuahua is a dry, bare mountain range called Sierra de Naica. In 2000, miners worked their way down through a network of caves below the mountains. When they got a thousand feet underground, they found themselves in a place that seemed to belong to another world. They were standing in a chamber measuring thirty feet wide and ninety feet long. The ceiling, walls, and floor were lined with smooth-faced, translucent crystals of gypsum. Many caves contain crystals, but not like the ones in Sierra de Naica. They measured up to thirty-six feet long apiece and weighed as much as fifty-five tons. These were not crystals to hang from a necklace. These were crystals to climb like hills.

Since its discovery, a few scientists have been granted permission to visit this extraordinary chamber, known now as the Cave of Crystals. Juan Manuel García-Ruiz, a geologist at the University of Granada, was one of them. On the basis of his research, he was able to determine the age of the crystals. They formed 26 million years ago, when volcanoes began to form the mountains. Subterranean chambers took shape inside the mountains and filled with hot mineral-laced water. The heat of the volcanic magma kept the water heated to a scalding 136 degrees F, the ideal temperature for the minerals to settle out of the water and form crystals. Somehow the water stayed at that perfect temperature for hundreds of thousands of years, allowing the crystals to grow to surreal sizes.

In 2009, another scientist, Curtis Suttle, paid a visit to the Cave of Crystals. Suttle and his colleagues scooped up water from the chamber's pools and brought it back to their laboratory at the University of British Columbia to analyze. When you consider Suttle's line of work, his journey might seem like a fool's errand. Suttle has no professional interest in crystals, or minerals, or any rocks at all for that matter. He studies viruses.

There are no people in the Cave of Crystals for the viruses to infect. There are not even any fish. The cave has been effectively cut off from the biology of the outside world for millions of years. Yet Suttle's trip was well worth the effort. After he prepared his samples of crystal water, he gazed at them under a microscope. He saw viruses—swarms of them. There are as many as 200 million viruses in every drop of water from the Cave of Crystals.

That same year, another scientists named Dana Willner led a virus-hunting expedition of her own. Instead of a cave, she dove into the human body. Willner had people cough up sputum into a cup, and out of that fluid she and her colleagues fished out fragments of DNA. They compared the DNA fragments to millions of sequences stored in online databases. Much of the DNA was human, but many fragments came from viruses. Before Willner's expedition, scientists had assumed the lungs of healthy people were sterile. Yet Willner discovered that, on average, people have 174 species of viruses in the lungs. Only 10 percent of the species Willner found bore any close kinship to any virus ever found before.

The other 90 percent were as strange as anything lurking in the Cave of Crystals.

Just about wherever scientists look—deep within the Earth, on grains of sand blown off the Sahara, in hidden lakes a mile below the Antarctic ice—they are discovering viruses faster than they can make sense of them. And the science of virology is still young. For thousands of years, we knew viruses only from their effects in sickness and death. Until recently, we did not know how to join those effects to their cause.

The very word *virus* began as a contradiction. We inherited the word from the Roman Empire, where it meant, at once, the venom of a snake or the semen of a man. Creation and destruction in one word.

Over the centuries, *virus* took on another meaning: it signified any contagious substance that could spread disease. It might be a fluid, like the discharge from a sore. It might be a substance that traveled mysteriously through the air. It might even impregnate a piece of paper, spreading disease with the touch of a finger.

Virus began to take on its modern meaning only in the late 1800s, thanks to an agricultural catastrophe. In the Netherlands, tobacco farms were swept by a disease that left plants stunted, their leaves a mosaic of dead and live patches of tissue. Entire farms had to be abandoned.

In 1879, Dutch farmers came to Adolph Mayer, a young agricultural chemist, to beg for help. Mayer carefully studied the scourge, which he dubbed tobacco mosaic disease. He investigated the environment in which the plants grew—the soil, the temperature, the sunlight. He could find nothing to distinguish the healthy plants from the sick ones. Perhaps, he thought, the plants were suffering from an invisible infection. Plant scientists had already demonstrated that fungi could infect potatoes and other plants, so Mayer looked for fungus on the tobacco plants. He found none. He looked for parasitic worms that might be infesting the leaves. Nothing.

Finally Mayer extracted the sap from sick plants and injected drops into healthy tobacco. The healthy plants turned sick. Mayer realized that some microscopic pathogen must be multiplying inside the plants. He took sap from sick plants and incubated it in his laboratory. Colonies of bacteria began to grow. They became large enough that Mayer could see them with his naked eye. Mayer applied these bacteria to healthy plants, wondering if they would trigger tobacco mosaic disease. They did nothing of the sort. And with that failure, Mayer's research ground to a halt. The world of viruses remained unopened.

A few years later, another Dutch scientist named Martinus Beijerinck picked up where Mayer left off. He wondered if something other than bacteria was responsible for tobacco mosaic disease—something far smaller. He ground up diseased plants and passed the fluid through a fine filter that blocked both plant cells and bacteria. When he injected the clear fluid into healthy plants, they became sick.

Beijerinck filtered the juice from the newly infected plants and found that he could infect still more tobacco. Something in the sap of the infected plants—something smaller than bacteria— could replicate itself and could spread disease. In 1898, Beijerinck called it a "contagious living fluid."

Whatever that contagious living fluid carried was different from any other kind of life biologists knew about. It was not only inconceivably small but also remarkably tough. Beijerinck could add alcohol to the filtered fluid, and it would remain infective. Heating the fluid to near boiling did it no harm. Beijerinck soaked filter paper in the infectious sap and let it dry. Three months later, he could dip the paper in water and use the solution to sicken new plants.

Beijerinck used the word *virus* to describe the mysterious agent in his contagious living fluid. It was the first time anyone used the word the way we do today. But in a sense, Beijerinck simply used it to define viruses by what they were *not*. They were not animals, plants, fungi, or bacteria. What exactly they were, Beijerinck could not say.

It soon became clear that what Beijerinck had discovered was just one kind of virus among many. In the early 1900s, other scientists used the same method of filters and infections to trace other diseases to other viruses. Eventually, they learned how to cultivate some viruses outside of living animals and plants, using nothing more than colonies of cells growing in dishes or flasks.

Yet these scientists *still* couldn't agree about what viruses actually were. Some argued viruses were simple chemicals. Others thought viruses were parasites that grew inside cells. The confusion over viruses was so profound that scientists could not even agree if viruses were living or dead. In 1923, the British virologist Frederick Twort declared, "It is impossible to define their nature."

That confusion began to disperse with the work of a chemist named Wendell Stanley. As a chemistry student in the 1920s, Stanley learned how to combine molecules into repeating patterns, forming crystals. Crystals could reveal things about substances that they otherwise kept secret. Scientists could shoot X-rays at the crystals, for example, and observe the directions that the reflected rays bounced away. The patterns produced by the X-rays offered clues to the molecules inside the crystals.

In the early 1900s, crystals helped solved one of biology's biggest mysteries: what enzymes are made of. Scientists had long known that enzymes were produced by animals and other living things to carry out different jobs, such as breaking down food. By making enzyme crystals, scientists discovered they are made of protein. Stanley wondered if viruses were proteins as well.

To find out, he starting trying to make crystals out of viruses. He chose a familiar species for his attempt: tobacco mosaic virus. Stanley collected the juice of infected tobacco plants and then passed it through fine filters, as Beijerinck had done four decades earlier. To allow the viruses to crystallize in pure form, Stanley tried to remove every type of compound from that contagious living fluid except for proteins.

After he had prepared his purified concoction, Stanley watched tiny needles form inside it. They grew into opalescent sheets. For the first time in history, a person could see viruses with the naked eye.

These virus crystals were at once as rugged as a mineral and as alive as a microbe. Stanley could store them away for months like table salt in a pantry. When he then added the crystals to water, they turned back into invisible viruses that could infect tobacco plants as viciously as before.

Stanley's experiment, which he published in 1935, dazzled the world. "The old distinction between death and life loses some of its validity," declared the *New York Times*.

But Stanley had also made a small but profound mistake. The British scientists Norman Pirie and Fred Bawden discovered in 1936 that viruses were not pure protein, but only 95 percent. The other 5 percent consisted of another molecule, a mysterious strand-shaped substance called nucleic acid. Nucleic acids, scientists would later discover, are the stuff of genes, the instructions for building proteins and other molecules. Our cells store their genes in double-stranded nucleic acids, known as deoxyribonucleic acid, or DNA for short. Many viruses have DNA-based genes as well. Other viruses, such as tobacco mosaic virus, have a single-stranded form of nucleic acids, called ribonucleic acid, or RNA.

Four years after Stanley crystallized tobacco mosaic viruses, a team of German scientists finally saw the individual viruses themselves. In the 1930s, engineers invented a new generation of microscopes able to see objects far smaller than had ever been seen before. Gustav Kausche, Edgar Pfannkuch, and Helmut Ruska mixed crystals of tobacco mosaic viruses into drops of purified water and put them under one of the new devices. In 1939, they reported that they could see minuscule rods, measuring about 300 nanometers long. No one had ever seen a living thing anywhere near so small. To contemplate the size of viruses, tap out a single grain of salt onto a table. Stare at the tiny cube. You could line up about ten skin cells along one side of it. You could line up about a hundred bacteria. And you could line up a thousand tobacco mosaic viruses, end to end, alongside that same grain of salt.

In the decades that followed, virologists went on to dissect viruses, to map their molecular geography. While viruses contain nucleic acids and proteins like our own cells, scientists found that differences between the structures of viruses and cells are many. A human cell is stuffed with millions of different molecules that it uses to sense its surroundings, crawl around, take in food, grow, and decide whether to divide in two or kill itself for the good of its fellow cells. Virologists found that viruses, as a rule, were far simpler. They typically were just protein shells holding a few genes.

Virologists discovered that viruses can replicate themselves, despite their paltry genetic instructions, by hijacking other forms of life. They inject their genes and proteins into a host cell, which they manipulate into producing new copies of themselves. One virus goes into a cell, and within a day a thousand viruses may come out.

By the 1950s, virologists had grasped these fundamental facts. But that understanding did not bring virology to a halt. For one thing, virologists knew little about the many different ways in which viruses make us sick. They didn't know why papillomaviruses can cause horns to grow on rabbits and cause hundreds of thousands of cases of cervical cancer each year. They didn't know what made some viruses deadly and others relatively harmless. They had yet to learn how viruses evade the defenses of their hosts and how they evolve faster than anything else on the planet. In the 1950s they did not know that a virus that would later be named HIV had already spread from chimpanzees and gorillas into our own species, or that thirty years later it would become one of the greatest killers in history. They could not have dreamed of the vast number of viruses that exist on Earth; they could not have guessed that much of life's genetic diversity is carried in viruses. They did not know that viruses help produce much of the oxygen we breathe and help control the planet's thermostat. And they certainly would not have guessed that the human genome is partly composed from thousands of viruses that infected our distant ancestors, or that life as we know it may have gotten its start four billion years ago from viruses.

Now scientists know these things—or, to be more precise, they know *of* these things. They now recognize that from the Cave of Crystals to the inner world of the human body, Earth is a planet of viruses. Their understanding is still rough, but it is a start.

So let us start as well.

OLD COMPANIONS

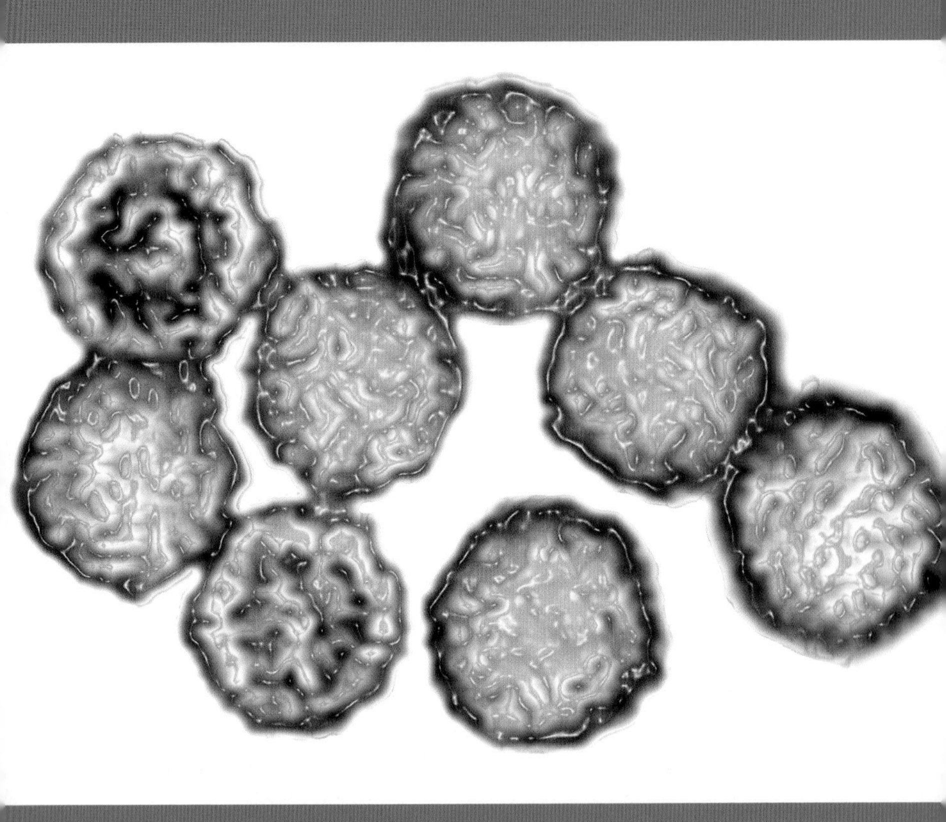

Rhinoviruses, the most common cause of colds

The Uncommon Cold

How Rhinoviruses Gently Conquered the World

Around 3,500 years ago, an Egyptian scholar sat down and wrote the oldest known medical text. Among the diseases he described in the so-called Ebers Papyrus was something called *resh*. Even with that strange-sounding name, its symptoms—a cough and a flowing of mucus from the nose—are immediately familiar to us all. *Resh* is the common cold.

Some viruses are new to humanity. Other viruses are obscure and exotic. But human rhinoviruses—the chief cause of the common cold, as well as asthma attacks—are old, cosmopolitan companions. It's been estimated that every human being will spend a year of his or her life lying in bed, sick with colds. The human rhinovirus is, in other words, one of the most successful viruses of all.

Hippocrates, the ancient Greek physician, believed that colds were caused by an imbalance of the humors. Two thousand years later, in the early 1900s, our knowledge of colds hadn't improved much. The physiologist Leonard Hill declared that colds were caused by walking outside in the morning, moving from warm air to cold.

In 1914, a German microbiologist named Walther Kruse gained the first solid clue about the origin of colds by having a snuffly assistant blow his nose. Kruse mixed the assistant's mucus into a salt solution, poured it through a filter, and then put a few drops of the filtered fluid into the noses of twelve colleagues. Four of them came down with colds. Later, Kruse did the same thing to thirty-six students. Fifteen of them got sick. Kruse compared their outcomes to thirty-five people who didn't get the drops. Only one of the drop-free individuals came down with a cold. Kruse's experiments made it clear that some tiny pathogen was responsible for the disease.

At first, many experts believed it was some kind of bacteria. But the American physician Alphonse Dochez ruled that out in 1927. He filtered the mucus from people with colds, using fine filters much as Beijerinck had filtered tobacco plant sap thirty years before. Even with the bacteria removed, the fluid could still make people sick. Only a virus could have slipped through Dochez's filters.

It took another three decades before scientists figured out exactly which viruses had slipped through. The most common of them are known as human rhinoviruses (*rhino* means nose). Rhinoviruses are remarkably simple, with only ten genes apiece. (Humans have about twenty thousand genes.) And yet their haiku of genetic information is enough to let rhinoviruses invade our bodies, outwit our immune system, and produce new viruses that can escape to new hosts.

Rhinoviruses spread by making noses run. People with colds wipe their noses, get the virus on their hands, and then spread the virus onto doorknobs and other surfaces they touch. The virus hitches onto the skin of other people who touch those surfaces and then slips into their bodies, usually through the nose. Rhinoviruses can invade the cells that line the interior of the nose, throat, or lungs. They trigger the cells to open up a trapdoor through which they slip. Over the next few hours, a rhinovirus will use its host cells to make copies of its genetic material and protein shells to hold them. The host cell then rips apart, and the new virus escapes.

Rhinoviruses infect relatively few cells, causing little real harm. So why can they cause such miserable experiences? We have only ourselves to blame. Infected cells release signaling molecules, called cytokines, which attract nearby immune cells. Those immune cells then make us feel awful. They create inflammation that triggers a scratchy feeling in the throat and leads to the production of a lot of mucus around the site of the infection. In order to recover from a cold, we have to wait not only for the immune system to wipe out the virus, but also for the immune system itself to calm down.

The Egyptian author of the Ebers Papyrus wrote that the cure for *resh* was to dab a mixture of honey, herbs, and incense around the nose. Fifteen centuries later, the Roman scholar Pliny the Elder recommended rubbing a mouse against the nose instead. In seventeenth-century England, cures included a blend of gunpowder and eggs and of fried cow dung and suet. Leonard Hill, the physiologist who believed a change of temperature caused colds, recommended that children start their day with a cold shower.

Today, there's still no cure for the common cold. The best treatment yet found is zinc, which blocks the growth of rhinoviruses. People who start taking zinc within a day of the start of a cold can shave off a day or more from their illness. Parents often give children cough syrup for colds, but studies show it doesn't make people get better faster. In fact, cough syrup also poses a wide variety of rare yet serious side effects, such as convulsions, rapid heart rate, and even death. The US Food and Drug Administration warns that children under the age of two—the people who get colds the most—should not take cough syrup.

All too often, doctors end up giving antibiotics to their patients with colds. This is a fundamentally pointless treatment, because antibiotics work only on bacteria and are useless against viruses. Doctors sometimes prescribe them because it's not clear whether a patient has a cold or a bacterial infection. In other cases, they may be responding to pressure from worried parents to do *something*. But antibiotics aren't just useless for colds. They're also a danger to us all, because they help foster the evolution of increasingly drug-resistant bacteria in our bodies and in the environment. Failing to treat their patients, doctors are actually raising the risk of other diseases for everyone.

One reason the cold remains so hard to treat may be that we've underestimated the rhinovirus. It exists in many forms, and scientists are only starting to get a true reckoning of its genetic diversity. By the end of the twentieth century, scientists had identified dozens of strains, which belonged to two great lineages, known as HRV-A and HRV-B. In 2006, Ian Lipkin and Thomas Briese of Columbia University were searching for the cause of flu-like symptoms in New Yorkers who did not carry the influenza virus. They discovered that a third of them carried a strain of human rhinovirus that was not closely related to either HRV-A or HRV-B. Lipkin and Briese dubbed it HRV-C. Since their discovery, researchers have found HRV-C all around the world. From one region to another, the variations in HRV-C's genes are few. Their uniformity suggests that this lineage emerged just a few centuries ago and rapidly spread around the world.

The more strains of rhinoviruses scientists discover, the better they come to understand their evolution. All human rhinoviruses share a core of genes that have changed very little over the centuries. Meanwhile, a few parts of the rhinovirus genome are evolving very quickly. These regions appear to help the virus avoid being killed by our immune systems. When our bodies build antibodies that can stop one strain of human rhinovirus, other strains can still infect us because our antibodies don't fit on their surface proteins. Consistent with this hypothesis is the fact that people are typically infected by several different human rhinovirus strains each year.

The diversity of human rhinoviruses makes them a very difficult target to hit. A drug or a vaccine that attacks one protein on the surface of one strain may prove to be useless against others that have a version of that protein with a different structure. If another strain of human rhinovirus is even a little resistant to such treatments, natural selection can foster the spread of new mutations, leading to much stronger resistance.

Despite this daunting diversity of rhinoviruses, some scientists still think it may be possible to create a cure for the common cold. The fact that all strains of human rhinoviruses share a common core of genes suggests that the core can't withstand mutations. If scientists can figure out ways to attack the rhinovirus's genetic core, they may be able to stop the disease.

One promising target in the rhinovirus core is a stretch of genetic material that folds into a loop shaped like a cloverleaf. Every rhinovirus scientists have studied carries the same cloverleaf structure. It appears to be essential for speeding up the rate at which a host cell copies rhinovirus genes. If scientists can find a way to disable the cloverleaf, they may be able to stop every cold virus on Earth.

But should they? The answer is actually not clear. Human rhinoviruses impose a serious burden on public health, not just by causing colds, but by opening the way for more harmful pathogens. Yet the effects of human rhinovirus itself are relatively mild. Most colds finish in under a week, and 40 percent of people who test positive for rhinoviruses suffer no symptoms at all. In fact, human rhinoviruses may offer some benefits to their human hosts. Scientists have gathered a great deal of evidence that children who get sick with relatively harmless viruses and bacteria may be protected from immune disorders when they get older, such as allergies and Crohn's disease. Human rhinoviruses may help train our immune systems not to overreact to minor triggers, instead directing their assaults to real threats. Perhaps we should not think of colds as ancient enemies but as wise old tutors.

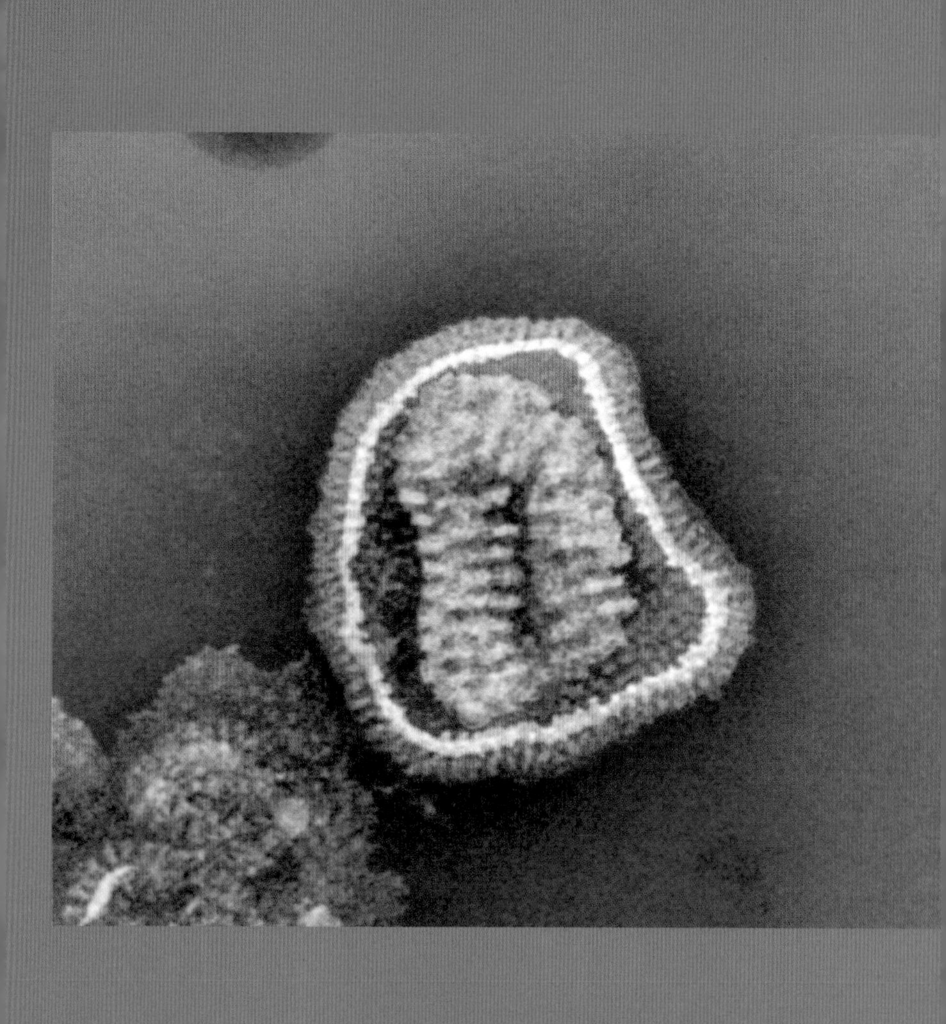

Influenza virus: the envelope layer appears orange and capsid is gray-white, with purple RNA segments inside

Looking Down from the Stars

Influenza's Never-Ending Reinvention

Influenza. If you close your eyes and say the word aloud, it sounds lovely. It would make a good name for a pleasant, ancient Italian village. *Influenza* is, in fact, Italian (it means influence). It is also, in fact, an ancient name, dating back to the Middle Ages. But the charming associations stop there. Medieval physicians believed that stars influenced the health of their patients, sometimes causing a mysterious fever that swept across Europe every few decades. Influenza has continued to burden the world with periodic devastation. In 1918, a particularly virulent outbreak of the flu infected 500 million people—a third of humanity at the time—and killed an estimated 50 million people. Even in years without an epidemic, influenza takes a brutal toll. The World Health Organization estimates that each year the flu infects 5 to 10 percent of all adults and 20 to 30 percent of all children. Somewhere between a quarter and half a million people die of the flu each year.

Today scientists know that influenza is the work not of the heavens, but of a microscopic virus. Like cold-causing rhinoviruses, influenza viruses manage to wreak their harm with very little genetic information—just thirteen genes. They spread in the droplets sick people release with their coughs, sneezes, and runny noses. A new victim may accidentally breathe in a virus-laden droplet or pick it up on a doorknob and then bring now-contaminated fingers in contact with the mouth. Once a flu virus gets into the nose or throat, it can latch onto a cell lining the airway and slip inside. As flu viruses spread from cell to cell in the lining of the airway, they leave destruction in their wake. The mucus and cells lining the airway get destroyed, as if the flu viruses were a lawn mower cutting grass.

When healthy people get infected by influenza viruses, their immune systems can launch a counterattack in a matter of days. In such cases, the flu causes a wave of aches, fevers, and fatigue, but the worst of it is over within a week. In a small fraction of its victims, the flu virus opens the way for more serious infections. Normally, the top layer of cells serves as a barrier against a wide array of pathogens. The pathogens get trapped in the mucus, and the cells snag them with hairs, swiftly notifying the immune system of intruders. Once the influenza lawnmower has cut away that protective layer, pathogens can slip in and cause dangerous lung infections, some of which can be fatal.

There are still paradoxes to influenza's effects that virologists don't understand. Seasonal flu is most dangerous for people with weak immune systems—particularly young children and the elderly—because they can't keep the virus in check. But in the 1918 outbreak, it was young adults—people with strong immune systems—who proved to be particularly vulnerable. One theory holds that certain strains of the flu provoke the immune system to respond so aggressively that it ends up devastating the host instead of wiping out the virus. But some scientists doubt this explanation and think the true answer lies elsewhere. It's possible, for example, that in 1918, older people carried protective antibodies from a similar pandemic in 1889.

While the effects of the flu may still be mysterious, its origins are clear. It came from birds. Birds carry all known strains of human influenza viruses, along with a vast diversity of other flu viruses that don't infect humans. Many birds carry the flu without getting sick. Rather than infecting their airways, flu viruses typically infect the guts of birds; the viruses are then shed in bird droppings. Healthy birds that ingest the virus-laden water become infected in turn.

Sometimes a strain of bird flu will end up in people. They may work on a chicken farm, or butcher poultry in a market. A bird flu virus that ends up in a human airway may seem out of place. But it turns out the receptors on cells in bird guts are similar in shape to those in our airways. Bird flu viruses can sometimes latch onto those receptors and slip inside.

But the transition from bird to human is not a simple one. The genes a bird flu virus needs to thrive are different from those needed inside a human body. Human bodies are cooler than bird bodies, for example, and that difference means that molecules need different shapes to run efficiently.

As a result, bird flu viruses that leap into humans usually wind up in viral dead ends because they can't spread from person to person. Starting in 2005, for example, a strain of flu from birds called H5N1 began to sicken hundreds of people in Southeast Asia. It proved to be much deadlier than ordinary strains of seasonal flu, and so public health workers tracked it closely, taking measures to halt its spread. But year in and year out, it never showed any ability to move from one human to another. H5N1 viruses that wound up in people always died out—either because their hosts destroyed them, or because they killed their hosts.

But every now and then, bird flu viruses adapt to our bodies. Each time they replicate, the new viruses contain genetic mistakes known as mutations. Some of the mutations have no effect on viruses. Some leave them unable to reproduce. But a few mutations give flu viruses a reproductive advantage.

Natural selection favors these beneficial mutations. Some help the virus by altering the shape of the proteins that stud the virus shell, allowing them to grab human cells more effectively. Others help them spread from person to person.

Once a flu strain gets established in humans, it can spread around the world and begin to pulsate in a seasonal rise and fall. In places like the United States, most flu cases occur during the winter. According to one hypothesis, this is because the air is dry enough in those months to allow virus-laden droplets to float in the air for hours, increasing their chances of encountering a new host. At other times of the year, the humidity causes the droplets to swell and fall to the ground.

When a flu virus hitches a ride aboard a droplet and infects a new host, it sometimes invades a cell that's already harboring another flu virus. And when two different flu viruses reproduce inside the same cell, things can get messy. The genes of a flu virus are stored on eight separate segments, and when a host cell starts manufacturing the segments from two different viruses at once, they sometimes get mixed together. The new offspring end up carrying genetic material from both viruses. This mixing, known as reassortment, is a viral version of sex. When humans have children, the parents' genes are mixed together, creating new combinations of the same two sets of DNA. Reassortment allows flu viruses to mix genes together into new combinations of their own.

A quarter of all birds with the flu have two or more virus strains inside them at once. The viruses swap genes and gain new adaptations, such as the ability to move from living in wild birds to chickens, or even to mammals such as horses or pigs. And sometimes, on very rare occasions, reassortment can combine genes from avian and human viruses, creating a recipe for disaster. The new strain that results from this combination can easily spread from person to person. And because it has never circulated among humans before, no one has any defenses that could slow the new strain's spread.

Once bird flu viruses evolve into human pathogens, they continue to swap genes among themselves. This ongoing reassortment allows the viruses to escape destruction. Before people's immune systems get too familiar with a flu strain's surface proteins, it can use a little viral sex to take on a new disguise.

The role that reassortment played in the most recent pandemic was especially complex. The new strain, called Human/Swine 2009 H1N1, first surfaced in Mexico in March 2009. But it was decades in the making.

By sequencing its genes, scientists have traced the virus's origins to four separate strains. The oldest of these strains has been infecting pigs since the 1918 pandemic (it's possible the pigs got it from us). A second strain emerged in the 1970s, when a bird flu infected pigs in either Europe or Asia. And a third strain jumped later from birds to pigs in the United States. By the late 1990s, all three strains had combined into a new form, called a "triple reassortant" by scientists. Unnoticed at the time, it silently moved from pig to pig in large closed barns. The triple reassortant then combined with another pig flu virus, and was then able to jump into humans. It's likely that Human/Swine 2009 H1N1 made the jump in the fall of 2008. It sickened people for months before coming to light in the spring of 2009.

As soon as public health workers recognized that they had a new pandemic on their hands, they launched a global campaign to protect people from infection. Despite their efforts, Human/Swine 2009 H1N1 was able to sweep around the world, much like pandemics before it, infecting 10 to 20 percent of all people on Earth. The United States scrambled to make a new flu vaccine for Human/Swine 2009 H1N1, but the vaccine wasn't ready until the fall, and in the end it provided only moderate protection against the virus. If Human/Swine 2009 H1N1 had been as lethal as the 1918 flu, we might have faced another gigantic catastrophe with millions dead. In the end, it proved milder, causing about 250,000 deaths before ebbing away.

As I write in 2015, scientists are looking at the next potential pandemics. It may take only a few mutations for a strain of bird flu to evolve into a new strain of human influenza virus. Reassortment could accelerate the change. No one can say when, or if, any particular strain will make the jump. But we are not helpless as we wait to see what evolution has in store for us. We can do things to slow the spread of the flu, such as washing our hands. And scientists are learning how to make more effective vaccines by tracking the evolution of the flu virus so they can do a better job of predicting which strains will be most dangerous in flu seasons to come. We may not have the upper hand over the flu yet, but at least we no longer have to look to the stars to defend ourselves.

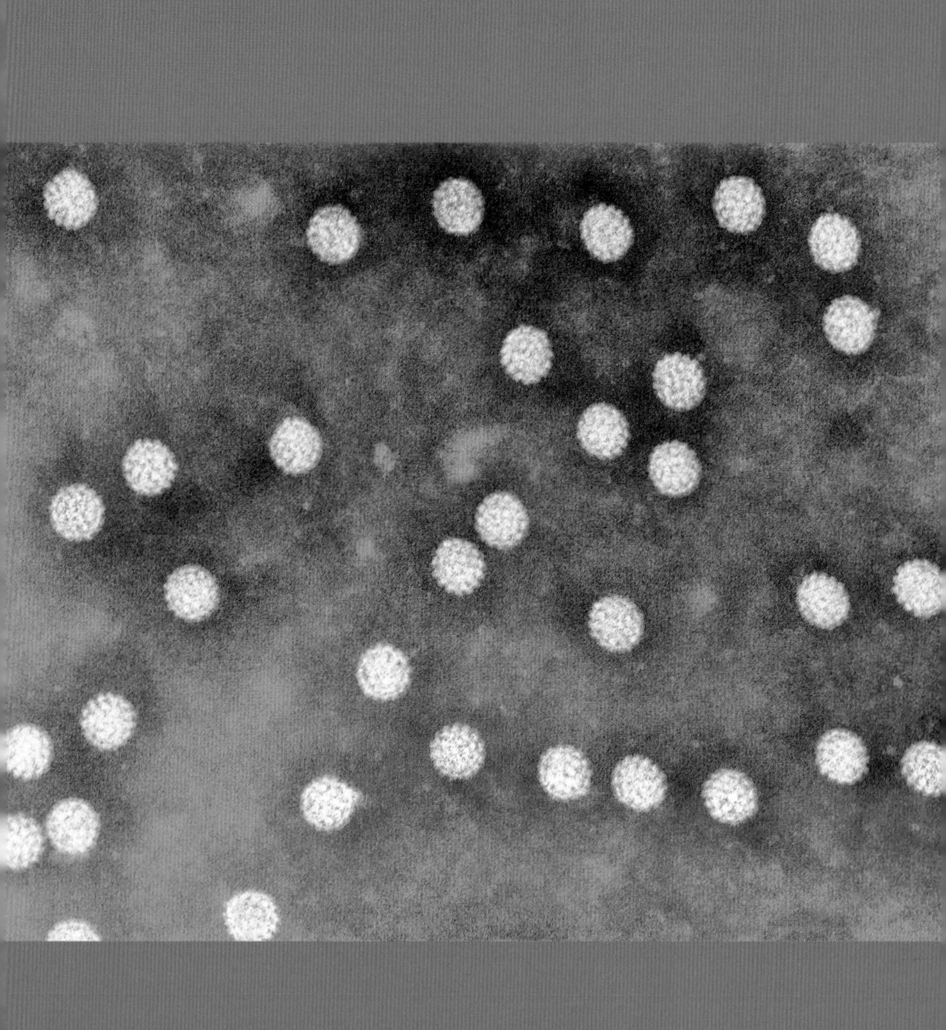

Human papillomaviruses (HPV) in suspension

Rabbits with Horns

Human Papillomavirus and Infectious Cancer

The stories about rabbits with horns circulated for centuries. Eventually they crystallized into the myth of the jackalope. If you go to Wyoming and twirl a rack of postcards, chances are you'll find a picture of a jackalope bounding across the prairie. It looks like a rabbit sprouting a pair of antlers. You may even see jackalopes in the flesh—or at least the head of one mounted on a diner wall.

On one level, it's all bunk. Most jackalopes are nothing but taxidermic trickery—rabbits with pieces of antelope antler glued to their heads. But like many myths, the tale of the jackalope has a grain of truth buried at its core. Some real rabbits do indeed sprout horn-shaped growths from their heads.

In the early 1930s, Richard Shope, a scientist at Rockefeller University, heard about horned rabbits while on a hunting trip. He had a friend catch one and send him some of the tissue so that he could figure out what it was made of. Shope's colleague, Francis Rous, had done experiments with chickens that suggested viruses could cause tumors. Many scientists at the time were skeptical, but Shope wondered if rabbit "horns" were also tumors, somehow triggered by an unknown virus. To test his hypothesis, Shope ground up the horns, mixed them in a solution, and then filtered the liquid through porcelain. The fine pores of the porcelain would let only viruses through. Shope then rubbed the filtered solution onto the heads of healthy rabbits. They grew horns as well.

Shope's experiment did more than show that the horns contained viruses. He also demonstrated that the viruses *created* the horns, crafting them out of infected cells. After this discovery, Shope passed on his rabbit tissue collection to Rous, who continued to work on it for decades. Rous injected virus-loaded liquid deep inside rabbits and found that it didn't produce harmless horns. Instead, the rabbits developed aggressive cancers that killed them. For his research linking viruses and cancer, Rous won the Nobel Prize in Medicine in 1966.

The discoveries of Shope and Rous led scientists to look at growths on other animals. Cows sometimes develop monstrous lumps of deformed skin as big as grapefruits. Warts grow on mammals, from dolphins to tigers to humans. And on rare occasions, warts can turn people into human jackalopes.

In the 1980s, a teenage boy in Indonesia named Dede began to develop warts on his body, and soon they had completely overgrown his hands and feet. Eventually he could no longer work at a regular job and ended up as an exhibit in a freak show, earning the nickname "Tree Man." Reports of Dede began to appear in the news, and in 2007 doctors removed thirteen pounds of warts from Dede's body. New growths returned, though, and so Dede had to undergo more surgery from time to time.

Dede's growths, along with all the others on humans and mammals, turned out to be caused by a single virus—the same kind that puts horns on rabbits. It's known as the papillomavirus, named for the papilla (*buds* in Latin) that infected cells form.

In the 1970s, the German researcher Harald zur Hausen speculated that papillomaviruses might be a threat to human health far bigger than the occasional wart. He wondered whether they might also cause tumors in the cervixes of women. Previous studies on cases of cervical cancer revealed patterns that were similar to sexually transmitted diseases. Nuns, for example, get cervical cancer much less often than other women. Some scientists had speculated that a virus spread during sex caused cervical cancer. Zur Hausen wondered if cancer-causing papillomaviruses were the culprit.

Zur Hausen reasoned that if this were true, he ought to find virus DNA in cervical tumors. He gathered biopsies to study, and slowly sorted through their DNA for years. In 1983 he discovered genetic material from papillomaviruses in the samples. As he continued to study the biopsies, he found more strains of papillomaviruses. Since zur Hausen first published his discoveries, scientists have identified one hundred different strains of human papillomavirus (or HPV for short). For his efforts, zur Hausen shared the Nobel Prize for Physiology or Medicine in 2008.

Zur Hausen's research put human papillomaviruses in medicine's spotlight, thanks to the huge toll that cervical cancer takes on the women of the world. The tumors caused by HPV grow so large that they sometimes rip the uterus or intestines apart. The bleeding can be fatal. Cervical cancer kills over 270,000 women every year, making it the third leading cause of death in women, surpassed only by breast cancer and lung cancer.

All of those cases got their start when a woman acquired an infection of HPV. The infection begins when the virus injects its DNA into a host cell. HPV specializes in infecting epithelial cells, which make up much of the skin and the body's mucous membranes. The virus's genes ends up inside the nucleus of its host cell, the home of the cell's own DNA. The cell then reads the HPV genes and makes the virus's proteins. Those proteins begin to alter the cell.

Many other viruses, such as rhinoviruses and influenza viruses, reproduce violently. They make new viruses as fast as possible, until the host cell brims with viral offspring. Ultimately, the cell rips open and dies. HPV uses a radically different strategy. Instead of killing its host cell, it causes the cell to make more copies of itself. The more host cells there are, the more viruses there are.

Speeding up a cell's division is no small feat, especially for a virus with just eight genes. The normal process of cell division is maddeningly complex. A cell "decides" to divide in response to signals both from the outside and the inside, mobilizing an army of molecules to reorganize its contents. Its internal skeleton of filaments reassembles itself, pulling apart the cell's contents to two ends. At the same time, the cell makes a new copy of its DNA—3.2 billion "letters" all told, organized into 46 clumps called chromosomes. The cell must drag those chromosomes to either end of the cell and build a wall through its center.

During this buzz of activity, supervising molecules monitor the progress. If the cell shows signs of becoming cancerous, they can trigger it to commit suicide. HPV can manipulate this complex dance by producing just a few proteins that intervene at crucial points in the cell cycle, accelerating it without killing the cell.

Many types of cells grow quickly in childhood and then slow down or stop altogether. Epithelial cells, the cells that HPV infects, continue to grow through our whole life. They start out in a layer buried below the skin's surface. As they divide, they produce a layer of new cells that pushes up on the cells above them. As the cells divide and rise, they become different than their progenitors. They begin to make more of a hard protein called keratin (the same stuff that makes up fingernails and horse hooves). Loaded with keratin, the top layer of skin can better withstand the damage from the sun, chemicals, and extreme temperatures. But eventually the top layer of epithelial cells dies off, and the next rising layers of epithelial cells take its place.

This arrangement means that HPV has to try to live on a conveyor belt. As HPV-infected cells reproduce, they move upward, closer and closer to their death. The viruses sense when their host cells are getting close to the surface and shift their strategy. Instead of speeding up cell division, they issue commands to their host cell to make many new viruses. When the cell reaches the surface, it bursts open with a big supply of HPV that can seek out new hosts to infect.

For most people infected with HPV, a peaceful balance emerges between virus and host. Fast-growing infected cells don't cause people harm, because they get sloughed off. The virus, meanwhile, gets to use epithelial cells as factories for new viruses, which can then infect new hosts through skin-to-skin contact and sex. The immune system helps maintain the balance by clearing away some of the infected cells. (Dede's tree-like growths were the result of a genetic defect that left his body unable to rein in the virus.)

This balance between host and virus has existed for hundreds of millions of years. To reconstruct the history of papillomaviruses, scientists compare the genetic sequence of different strains and note which animals they infect. It turns out that papillomaviruses infect not just mammals, such as humans, rabbits, and cows, but other vertebrates as well, such as birds and reptiles. Each strain of virus typically infects only one or a few related species. On the basis of their relationships, Marc Gottschling of the University of Munich has argued that the first egg-laying land vertebrates— the ancestor of mammals, reptiles, and birds—was already a host to papillomaviruses 300 million years ago.

As the descendants of that ancient animal evolved into different lineages, their papillomaviruses evolved as well. Some research suggests that these viruses began to specialize on different kinds of lining in their hosts. The viruses that cause warts, for example, adapted to infect skin cells. Another lineage adapted to the mucosal linings of the mouth and other orifices. For the most part, these new papillomaviruses coexisted peacefully with their hosts. Two-thirds of healthy horses carry strains of papillomavirus called BPV1 and BPV2. Some strains evolved to be more prone to turn cancerous than others, but researchers can't say why.

For thousands of generations, papillomaviruses have specialized on certain hosts, but from time to time, they have leapt to new species. A number of human papillomaviruses are most closely related to papillomaviruses that infect distantly related animals, like horses, instead of our closest ape relatives. Nothing more than skin contact may have been enough to allow viruses to make the jump.

When our own species first evolved in Africa about two hundred thousand years ago, our ancestors probably carried several different strains of papillomaviruses. Representatives of those strains can be found all over the world. But as humans expanded across the planet—leaving Africa about fifty thousand years ago and reaching the New World by about fifteen thousand years ago—their papillomaviruses were continuing to evolve. We know this because the genealogy of some HPV strains reflects the genealogy of our species. The viruses that infect living Africans belong to the oldest lineages of HPV, for example, while Europeans, Asians, and Native Americans carry their own distinct strains.

For about 199,950 of the past 200,000 years, our species had no idea that we were carrying HPV. That's not because HPV was rare—far from it: a 2014 study on 103 healthy people detected the viruses in 71 of them—about 69 percent. For the overwhelming majority of those people, the experience was harmless. Of the estimated 30 million American women who carry HPV, only 13,000 a year develop cervical cancer.

In this cancer-stricken minority, the peaceful balance between host and virus is thrown off. Natalia Shulzhenko of Oregon State University and her colleagues have proposed that HPV triggers cancer when some of its genetic material gets accidentally incorporated into the DNA of a cell. The cell multiplies rapidly and picks up new mutations. Instead of dying off, it remains in an unnaturally youthful state. Instead of getting sloughed off, the cells form a tumor, which pushes out and down into the surrounding tissue.

The best way to prevent most cancers is to reduce the odds that our cells will pick up dangerous mutations: quitting smoking, avoiding cancer-promoting chemicals, and eating well. But cervical cancer can be blocked another way: with a vaccine. In 2006, the first HPV vaccines were approved for use in the United States and Europe. They all contain proteins from the outer shell of HPV, which the immune system can learn to recognize. If people are later infected with HPV, their immune systems can mount a rapid attack and wipe it out.

The introduction of the vaccine unleashed controversy in the United States, because the manufacturer, GlaxoSmithKline, recommended that children get vaccinated between ages eleven and thirteen. Some parents protested that such a policy promotes sex before marriage. Those concerns helped to keep vaccination rates low. In 2013, only 35 percent of boys and 57 percent of girls received the HPV vaccine by their thirteenth birthday. The Centers for Disease Control decried that rate as "unacceptably low." No matter what objections parents may have about the vaccine, it unquestionably works. Long-term studies have shown that it provides complete protection against the two strains that cause 70 percent of cervical cancer cases.

Even if all children get the vaccine, however, cervical cancer may not disappear altogether. At best, the two strains targeted by the vaccine could be eradicated. But scientists have identified thirteen other cancer-causing strains of HPV, and there are likely others yet to be discovered. If vaccines decimate the two most successful strains, natural selection might well favor the evolution of other strains to take their place. Never underestimate the evolutionary creativity of a virus that can transform rabbits into jackalopes or men into trees.

EVERYWHERE,
IN ALL THINGS

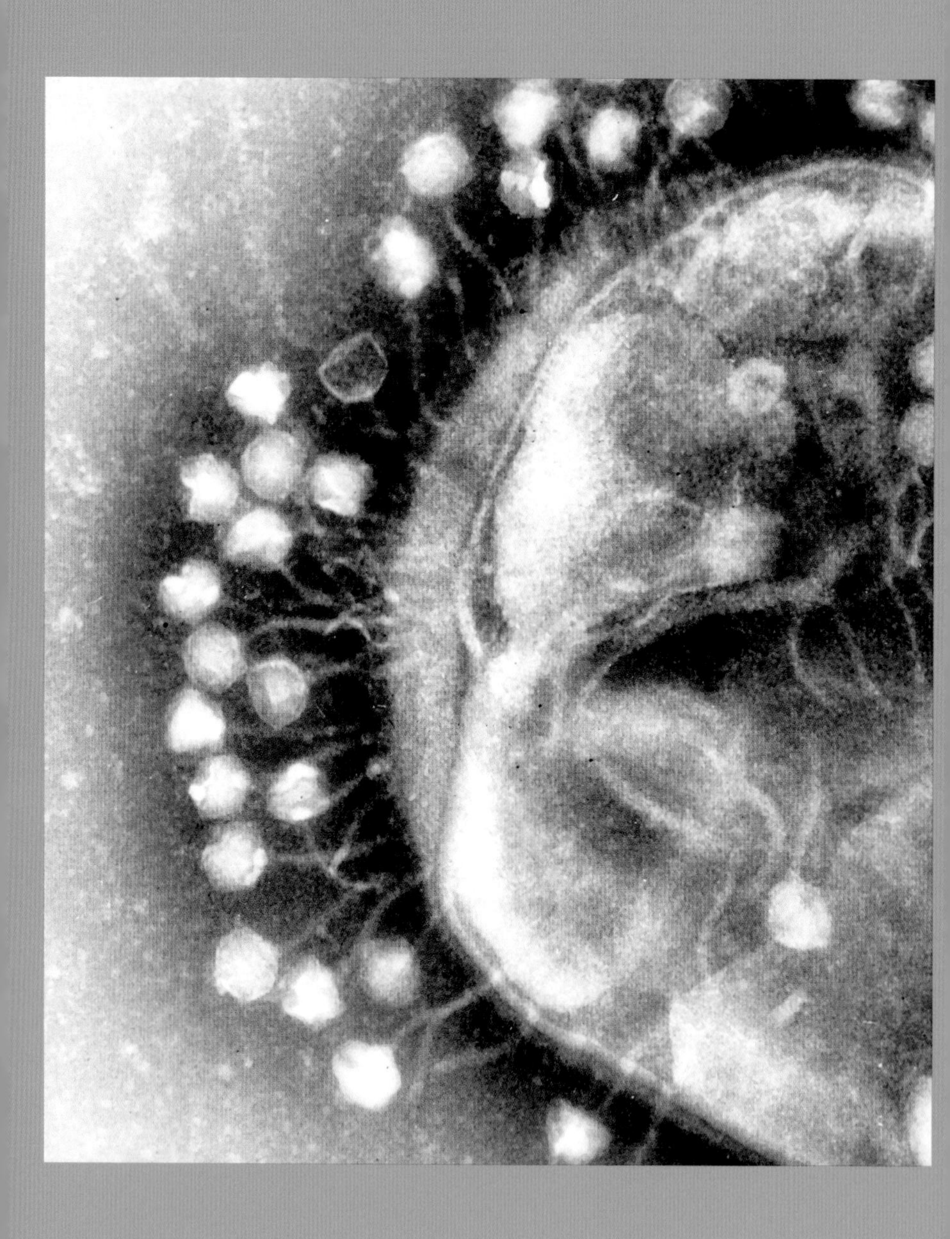

Bacteriophages attach to the surface of the host cell,
a bacterium *Escherichia coli*

The Enemy of Our Enemy

Bacteriophages as Viral Medicine

By the beginning of the twentieth century, scientists had learned a few important things about viruses. They knew that viruses were infectious agents of unimaginably small size. They knew that certain viruses caused certain diseases, such as tobacco mosaic disease and rabies. But the young science of virology was still parochial. It focused mainly on the viruses that worried people most: the ones that made people sick, and the ones that threatened the crops and livestock they grew. Virologists rarely looked beyond our little circle of experience. But during World War I, two physicians independently got a glimpse of the greater universe of viruses in which we live.

In 1915, an English physician named Frederick Twort discovered this universe quite by accident. At the time, he was looking for an easier way to make smallpox vaccines. In the early 1900s, the standard vaccine for the disease contained a mild relative of smallpox called vaccinia. Unfortunately, new vaccinia viruses could be collected only from hosts: either from immunized people or immunized animals. Twort wondered if it might be possible to grow vaccinia faster, by infecting cells he grew on lab plates.

His experiments ended in failure, because bacteria contaminated his plates and wiped out his cells. But Twort's frustration wasn't enough to blind him to something strange. He noticed glassy spots developing in his plates. Under a microscope, Twort could see that the spots were full of dead microbes. He collected tiny drops from the spots and transferred them to living bacterial colonies. In a matter of hours, glassy spots formed in the new dishes, full of more dead bacteria. But if he added drops to a different species of bacteria, no spots formed.

Twort could think of three explanations for what he was seeing. It might be some bizarre feature of the life cycle of the bacteria. Or perhaps the bacteria were making some enzyme that was killing them. The third possibility was the hardest to believe: maybe Twort had discovered a virus that kills bacteria.

Twort published his findings, listed the three possibilities, and left matters there. The radical idea of bacteria-infecting viruses apparently didn't infect his mind. But two years later, a Canadian-born doctor named Felix d'Herelle independently made the same discovery, and he was not so resistant.

In 1917, Herelle was working as a military doctor, caring for French soldiers dying of dysentery. Dysentery is caused by bacteria known as *Shigella*. Today, doctors can use antibiotics to kill bacteria, but those drugs would not be discovered until decades after the Great War. To better understand his enemy, Herelle examined the diarrhea produced by the sick soldiers.

As part of his analysis, he passed the stool of the soldiers through fine filters, trapping *Shigella* and any other bacteria they contained. Once Herelle had produced this clear, bacteria-free fluid, he then mixed it with a fresh sample of *Shigella* bacteria and then spread the mixture of bacteria and clear fluid in petri dishes. The *Shigella* began to grow, but within a few hours Herelle noticed clear spots starting to form in their colonies. Herelle drew samples from those spots and mixed them with *Shigella* again. More clear spots formed in the dishes. These spots, Herelle concluded, were battlegrounds where viruses were killing *Shigella* and leaving behind their translucent corpses.

Herelle believed his discovery was so radical that his viruses deserved a name of their own. He dubbed them *bacteriophages*, meaning "eaters of bacteria." Today, they're known as phages for short.

The concept of bacteria-infecting viruses was so odd that some scientists refused to believe it. Jules Bordet, a French immunologist who won the Nobel Prize in 1919, became Herelle's most outspoken critic after he failed to find phages of his own. Instead of *Shigella*, Bordet used a harmless strain of *Escherichia coli*. He poured *E. coli*–laden liquid through fine filters, and then mixed the filtered liquid with a second batch of *E. coli*. The second batch died, just as they had in Herelle's experiments. But then Bordet decided to see what would happen if he mixed the filtered liquid with the first batch of *E. coli*—that is, the one he had filtered in the first place. To his surprise, the first batch of *E. coli* was immune.

Bordet believed that his failure to kill the bacteria meant that the filtered fluid did not contain phages. Instead, he thought, it contained a protein produced by the first *E. coli*. The protein was toxic to other bacteria, but not to the ones that made it.

Herelle fought back against Bordet, Bordet counterattacked, and the debate raged for years. It wasn't until the 1940s that scientists finally found the visual proof that Herelle was right. Settling the debate required the invention of electron microscopes powerful enough to see the minuscule viruses. When they mixed bacteria-killing fluid with *E. coli* and put it under the microscopes, they saw that bacteria were attacked by phages. The phages had boxlike shells in which their genes were coiled, sitting atop a set of what looked like spider legs. The phages dropped onto the surface of *E. coli* like a lunar lander on the moon and then drilled into the microbe, squirting in their DNA.

As scientists got to know phages better, it became clear that the debate between Herelle and Bordet was just a case of apples and oranges. Phages do not belong to a single species, and different phage species behave differently toward their hosts. Herelle had found a vicious form, called a lytic phage, which kills its host as it multiplies. Bordet had found a more benevolent kind of virus, which came to be known as a temperate phage. Temperate phages treat bacteria much like human papillomaviruses treat our skin cells. When a temperate phage infects its host microbe, its host does not burst open with new phages. Instead, the temperate phage's genes are joined into the host's own DNA, and the host continues to grow and divide. It is as if the virus and its host become one.

Once in a while, however, the DNA of the temperate phage awakens. It commandeers the cell to make new phages, which burst out of the cell and invade new ones. And once a temperate phage is incorporated into a microbe, the host becomes immune from any further invasion. That's why Bordet couldn't kill his first batch of *E. coli* with the phage—it was already infected, and thus protected.

Herelle did not wait for the debate over phages to end before he began to use them to cure his patients. During World War I, he observed that as soldiers recovered from dysentery and other diseases, the levels of phages in their stool climbed. Herelle concluded that the phages were actually killing the bacteria. Perhaps, if he gave his patients extra phages, he could eliminate diseases even faster.

Before he could test this hypothesis, Herelle first needed to be sure phages were safe. So he swallowed some to see if they made him sick. He found that he could ingest phages, as he later wrote, "without detecting the slightest malaise." Herelle injected phages into his skin, again with no ill effects. Confident that phages were safe, Herelle began to give them to sick patients. He reported that they helped people recover from dysentery and cholera. When four passengers on a French ship in the Suez Canal came down with bubonic plague, Herelle gave them phages. All four victims recovered.

Herelle's cures made him even more famous than before. The American writer Sinclair Lewis made Herelle's radical research the basis of his 1925 best-selling novel *Arrowsmith*, which Hollywood turned into a movie in 1931. Meanwhile, Herelle developed phage-based drugs sold by the company that's now known as L'Oreal. People used his phages to treat skin wounds and to cure intestinal infections.

But by 1940, the phage craze had come to end. The idea of using live viruses as medicine had made many doctors uneasy. When antibiotics were discovered in the 1930s, those doctors responded far more enthusiastically, because antibiotics were not alive; they were just artificial chemicals and proteins produced by fungi and bacteria. Antibiotics were also staggeringly effective, often clearing infections in a few days. Pharmaceutical companies abandoned Herelle's phages and began to churn out antibiotics. With the success of antibiotics, investigating phage therapy seemed hardly worth the effort.

Yet Herelle's dream did not vanish entirely when he died in 1949. On a trip to the Soviet Union in the 1920s, he had met scientists who wanted to set up an entire institute for research on phage therapy. In 1923 he helped Soviet researchers establish the Eliava Institute of Bacteriophage, Microbiology, and Virology in Tbilisi, which is now the capital of the Republic of Georgia. At its peak, the institute employed 1,200 people to produce tons of phages a year. During World War II, the Soviet Union shipped phage powders and pills to the front lines, where they were dispensed to infected soldiers.

In 1963, the Eliava Institute ran the largest trial ever conducted to see how well phages actually worked in humans, enrolling 30,769 children in Tbilisi. Once a week, about half the children swallowed a pill that contained phages against *Shigella*. The other half of the children got a pill made of sugar. To minimize environmental influences as much as possible, the Eliava scientists gave the phage pills only to children who lived on one side of each street, and the sugar pills to the children who lived on the other side. The scientists followed the children for 109 days. Among the children who took the sugar pill, 6.7 out of every 1,000 got dysentery. Among the children who took the phage pill, that figure dropped to 1.8 per 1,000. In other words, taking phages caused a 3.8-fold decrease in a child's chance of getting sick.

Few people outside of Georgia heard about these striking results, thanks to the secrecy of the Soviet government. Only after the Soviet Union fell in 1989 did news start to trickle out. The reports have inspired a small but dedicated group of Western scientists to investigate phage therapy and to challenge the long-entrenched reluctance in the West to use them.

These phage champions argue that we should not be worried about using live viruses as medical treatments. After all, phages swarm inside many of the foods we eat, such as yogurt, pickles, and salami. Our bodies are packed with phages too, which is not surprising when you consider that we each carry about a hundred trillion bacteria—all promising hosts for various species of phages. Every day, those phages kill vast numbers of bacteria inside our bodies without ever harming our health.

Another concern that's been raised about phages is that their attack is too narrowly focused. Each species of phage can attack only one species of bacteria, while one antibiotic can kill off many different species at once. But it's clear now that phage therapy can treat a wide range of infections. Doctors just have to combine many phage species into a single cocktail. Scientists at the Eliava Institute have developed a dressing for wounds that is impregnated with half a dozen different phages, capable of killing the six most common kinds of bacteria that infect skin wounds.

Skeptics have also argued that even if scientists could design an effective phage therapy, evolution would soon render it useless. In the 1940s, the microbiologists Salvador Luria and Max Delbruck observed phage resistance evolving before their own eyes. When they laced a dish of *E. coli* with phages, most of the bacteria died, but a few clung to existence and then later multiplied into new colonies. Further research revealed that those survivors had acquired mutations that allowed them to resist the phages. The resistant bacteria then passed on their mutated genes to their descendants. Critics have argued that phage therapy would also foster the evolution of phage-resistant bacteria, allowing infections to rebound.

The advocates for phage therapy respond by pointing out that phages can evolve, too. As they replicate, they sometimes pick up mutations, and some of those mutations can give them new avenues for infecting resistant bacteria. Scientists can even help phages improve their attacks. They can search through collections of thousands of different phages to find the best weapon for any particular infection, for example. They can even tinker with phage DNA to create phages that can kill in new ways.

In 2008, James Collins, a biologist at Boston University, and Tim Lu of MIT published details of the first phage engineered to kill. Their new phage is especially effective because it's tailored to attack the rubbery sheets that bacteria embed themselves in, known as biofilms. Biofilm can foil antibiotics and phages alike, because they can't penetrate the tough goo and reach the bacteria inside. Collins and Lu searched through the scientific literature for a gene that might make phages better able to destroy biofilms. Bacteria themselves carry enzymes that they use to loosen up biofilms when it's time for them to break free and float away to colonize new habitats.

Collins and Lu synthesized a gene for one of these biofilm-dissolving enzymes and inserted it into a phage. They then tweaked the phage's DNA so that it would produce lots of the enzyme as soon as it entered a host microbe. When they unleashed it on biofilms of *E. coli*, the phages penetrated the microbes on the top of the biofilms and forced them to make both new phages and new enzymes. The infected microbes burst open, releasing enzymes that sliced open deeper layers of the biofilms, which the phages could infect. The engineered phages can wipe 99.997 percent of the *E. coli* in a biofilm, a kill rate that's about a hundred times better than ordinary phages.

While Collins and other scientists discover how to make phages even more effective, antibiotics are now losing their luster. Doctors are grappling with a growing number of bacteria that have evolved resistance to most of the antibiotics available today. Sometimes doctors have to rely on expensive, last-resort drugs that come with harsh side effects. And there's every reason to expect that bacteria will evolve to resist last-resort antibiotics as well. Scientists are scrambling to develop new antibiotics, but it can take over a decade to get a new drug from the lab to the marketplace. It may be hard to imagine a world before antibiotics, but now we must imagine a world where antibiotics are not the only weapon we use against bacteria. And now, a century after Herelle first encountered bacteriophages, these viruses may finally be ready to become a part of modern medicine.

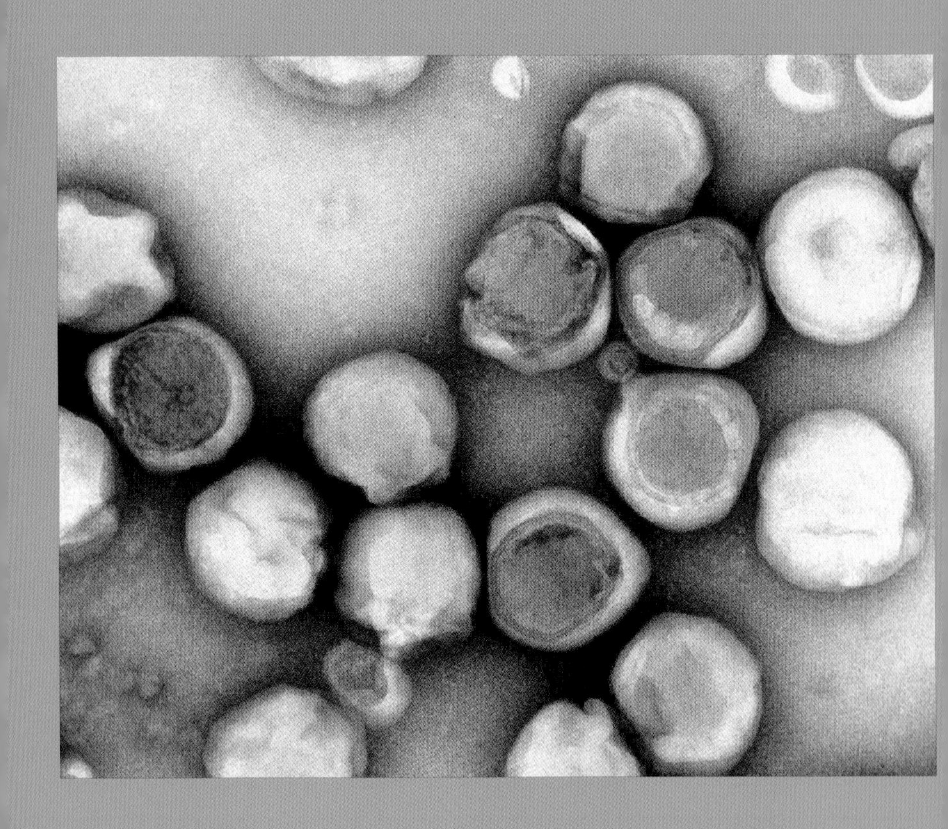

Emiliania huxleyi viruses infect ocean algae (viruses shown here in suspension)

The Infected Ocean

How Marine Phages Rule the Sea

Some great discoveries seem at first like terrible mistakes.

In 1986 a graduate student at the State University of New York at Stony Brook named Lita Proctor decided to see how many viruses there are in seawater. At the time, the general consensus was that there were hardly any. The few researchers who had bothered to look for viruses in the ocean had generally found only a scarce supply. Most experts believed that the majority of the viruses they did find in seawater had actually come from sewage and other sources on land.

But over the years, a handful of scientists had gathered evidence that didn't fit neatly into the consensus. A marine biologist named John Sieburth, for example, had published a photograph of a marine bacterium erupting with new viruses. Proctor decided it was time to launch a systematic search. She traveled to the Caribbean and to the Sargasso Sea, scooping up seawater along the way. Back on Long Island, she carefully extracted the biological material from the seawater, which she coated with metal so that it would show up under the beam of an electron microscope. When Procter finally looked at her samples, she beheld a world of viruses. Some floated freely, while others were lurking inside infected bacterial hosts. Based on the number of viruses she found in her samples, Proctor estimated that every liter of seawater contained up to one hundred billion viruses.

Proctor's figure far exceeded previous estimates. But when other scientists followed up on her work and carried out their own surveys, they ended up with similar figures. They came to agree that there are somewhere in the neighborhood of 10,000,000,000,000,000,000,000,000,000,000 viruses in the ocean.

It is hard to find a point of comparison to make sense of such a huge number. Viruses outnumber all other residents of the ocean by about fifteen to one. If you put all the viruses of the oceans on a scale, they would equal the weight of seventy-five million blue whales (there are less than ten thousand blue whales on the entire planet). And if you lined up all the viruses in the ocean end to end, they would stretch out 42 million light-years.

These numbers don't mean that a swim in the ocean is a death sentence. Only a minute fraction of the viruses in the ocean can infect humans. Some marine viruses infect fishes and other marine animals. But their most common targets by far are bacteria and other single-celled microbes. Microbes may be invisible to the naked eye, but collectively they dwarf all the ocean's whales, its coral reefs, and all other forms of marine life. And just as the bacteria that live in our bodies are attacked by phages, marine microbes are attacked by marine phages.

When Felix d'Herelle discovered the first bacteriophage in French soldiers in 1917, many scientists refused to believe that such a thing actually existed. A century later, it's clear that Herelle had found the most abundant life form on Earth. What's more, marine viruses have a massive influence on the planet. Marine phages influence the ecology of the world's oceans. They leave their mark on Earth's global climate. And they have been playing a crucial part in the evolution of life for billions of years. They are, in other words, biology's living matrix.

Marine viruses are powerful because they are so infectious. They invade a new microbe host ten trillion times a second, and they kill between 15 and 40 percent of all bacteria in the world's oceans every single day. By killing these hosts, they create swarms of new viruses. Every liter of seawater generates up to 100 billion new viruses every day—viruses that can promptly infect new hosts. Their lethal efficiency keeps their hosts in check, and we humans often benefit from their deadliness. Cholera, for example, is caused by blooms of waterborne bacteria called *Vibrio*. But *Vibrio* are host to a number of phages. When the population of *Vibrio* explodes and causes a cholera epidemic, the phages multiply. The virus population rises so quickly that it kills *Vibrio* faster than the microbes can reproduce. The bacterial boom subsides, and the cholera epidemic fades away.

Stopping cholera outbreaks is actually one of the smaller effects of marine viruses. They kill so many microbes that they can also influence the atmosphere across the planet. That's because microbes themselves are the planet's great geoengineers. Algae and photosynthetic bacteria churn out about half of the oxygen we breathe. Algae also release a gas called dimethyl sulfide that rises into the air and seeds clouds. The clouds reflect incoming sunlight back out into space, cooling the planet. Microbes also absorb and release vast amounts of carbon dioxide, which traps heat in the atmosphere. Some microbes release carbon dioxide into the atmosphere as waste, warming the planet. Algae and photosynthetic bacteria, on the other hand, suck carbon dioxide in as they grow, making the atmosphere cooler. When microbes in the ocean die, some of their carbon rains down to the sea floor. Over millions of years, this microbial snow can steadily make the planet cooler and cooler. What's more, these dead organisms can turn to rock. The White Cliffs of Dover, for example, are made up of the chalky shells of single-cell organisms called coccolithophores.

Viruses kill these geoengineers by the trillions every day. As the microbial victims die, they spill open and release a billion tons of carbon a day. Some of the liberated carbon acts as a fertilizer, stimulating the growth of other microbes, but some of it probably sinks to the bottom of the ocean. The molecules inside a cell are sticky, and so once a virus rips open a host, the sticky molecules that fall out may snag other carbon molecules and drag them down in a vast storm of underwater snow.

Ocean viruses are stunning not just for their sheer numbers but also for their genetic diversity. The genes in a human and the genes in a shark are quite similar—so similar that scientists can find a related counterpart in the shark genome to most genes in the human genome. The genetic makeup of marine viruses, on the other hand, matches almost nothing. In a survey of viruses in the Arctic Ocean, the Gulf of Mexico, Bermuda, and the northern Pacific, scientists identified 1.8 million viral genes. Only 10 percent of them showed any match to any gene from any microbe, animal, plant, or other organism—even from any other known virus. The other 90 percent were entirely new to science. In 200 liters of seawater, scientists typically find 5,000 genetically distinct kinds of viruses. In a kilogram of marine sediment, there may be a million kinds.

One reason for all this diversity is that marine viruses have so many hosts to infect. Each lineage of viruses has to evolve new adaptations to get past its host's defenses. But diversity can also evolve by more peaceful means. Temperate phages merge seamlessly into their host's DNA; when the host reproduces, it copies the virus's DNA along with its own. As long as a temperate phage's DNA remains intact, it can still break free from its host during times of stress. But over enough generations, a temperate phage will pick up mutations that hobble it, so that it can no longer escape. It becomes a permanent part of its host's genome.

As a host cell manufactures new viruses, it sometimes accidentally adds some of its own genes to them. The new viruses carry the genes of their hosts as they swim through the ocean, and they insert them, along with their own, into the genomes of their new hosts. By one estimate, viruses transfer a trillion trillion genes between host genomes in the ocean every year.

Sometimes these borrowed genes make the new host more successful at growing and reproducing. The success of the host means success for the virus, too. While some species of viruses kill *Vibrio*, others deliver genes for toxins that the bacteria use to trigger diarrhea during cholera infections. The new infection of toxin-carrying viruses may be responsible for new cholera outbreaks.

Thanks to gene borrowing, viruses may also be directly responsible for a lot of the world's oxygen. An abundant species of ocean bacteria, called *Synechococcus*, carries out about a quarter of the world's photosynthesis. When scientists examine the DNA of *Synechococcus* samples, they often find proteins from viruses carrying out their light harvesting. Scientists have even found free-floating viruses with photosynthesis genes, searching for a new host to infect. By one rough calculation, 10 percent of all the photosynthesis on Earth is carried out with virus genes. Breathe ten times, and one of those breaths comes to you courtesy of a virus.

This shuttling of genes has had a huge impact on the history of all life on Earth. It was in the oceans that life got its start, after all. The oldest traces of life are fossils of marine microbes dating back almost 3.5 billion years. It was in the oceans that multicellular organisms evolved; their oldest fossils date back to about 2 billion years ago. In fact, our own ancestors did not crawl onto land until about 400 million years ago. Viruses don't leave behind fossils in rocks, but they do leave marks on the genomes of their hosts. Those marks suggest that viruses have been around for billions of years.

Scientists can determine the history of genes by comparing the genomes of species that split from a common ancestor that lived long ago. Those comparisons can, for example, reveal genes that were delivered to their current host by a virus that lived in the distant past. Scientists have found that all living things have mosaics of genomes, with hundreds or thousands of genes imported by viruses. As far down as scientists can reach on the tree of life, viruses have been shuttling genes. Darwin may have envisioned the history of life as a tree. But the history of genes, at least among the ocean's microbes and their viruses, is more like a bustling trade network, its webs reaching back billions of years.

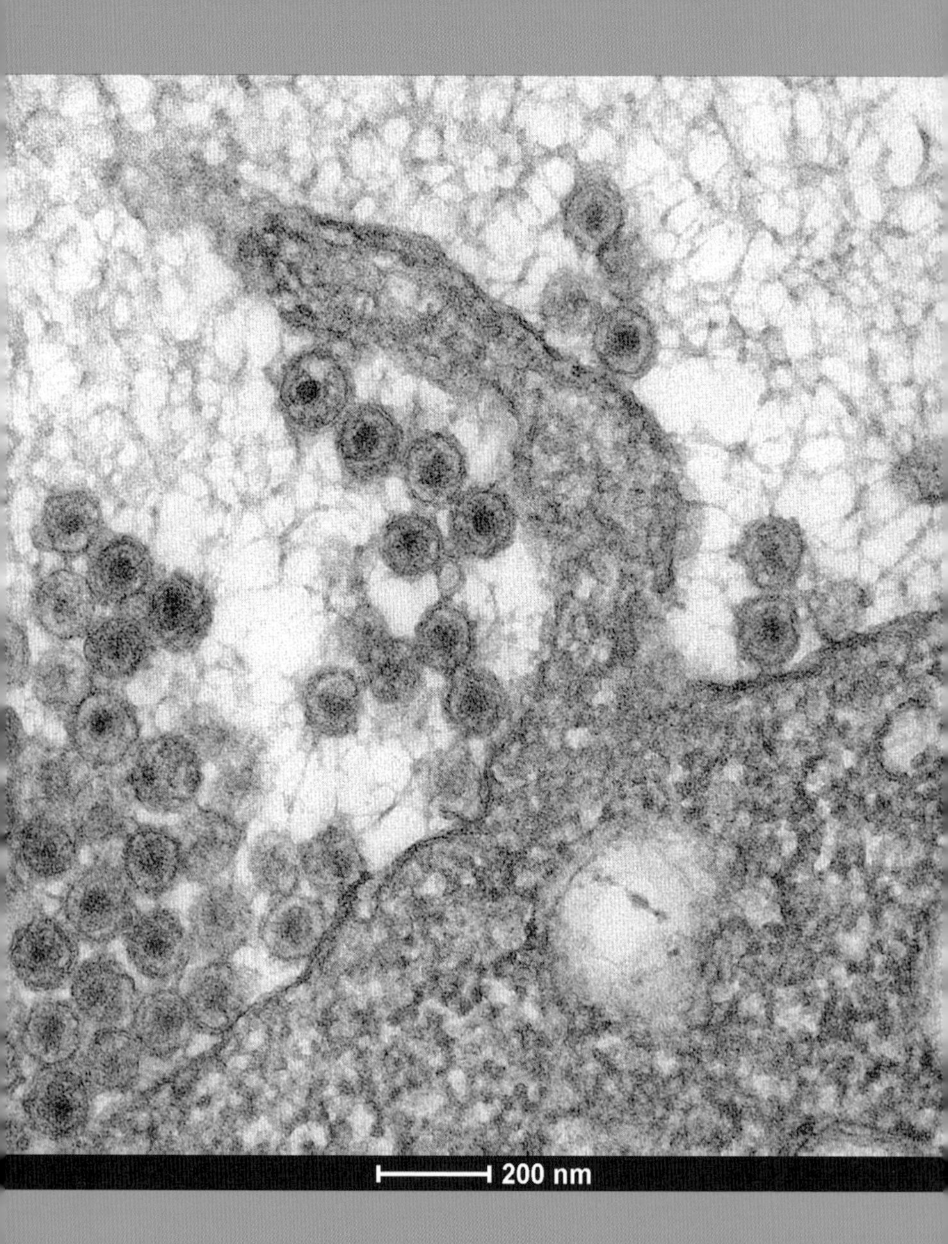

Avian leukocyte viruses bud from a human white blood cell

Our Inner Parasites

Endogenous Retroviruses and Our Virus-Riddled Genomes

The idea that a host's genes could have come from viruses is almost philosophical in its weirdness. We like to think of our genomes as our ultimate identity. The fact that bacteria have acquired much of their DNA from viruses raises baffling questions. Do they have a distinct identity of their own? Or are they just hybrid Frankensteins, their clear lines of identity blurred away?

At first, it was possible to cordon off this puzzle from our own existence, treating it purely as a question about microbes. The presence of viral genes was merely a fluke of "lower" life forms. But today we can no longer find such comfort. If we look inside our own genome, we now see viruses. Thousands of them.

We have the jackalope to thank for this realization. The myth of the jackalope was one of the clues that led virologists to discover that some viruses cause cancer. In the 1960s, one of the most intensely studied cancer-causing viruses was avian leukosis virus. At the time, the virus was sweeping across chicken farms and threatening the entire poultry industry. Scientists found that avian leukosis virus belonged to a group of species known as retroviruses. Retroviruses insert their genetic material into their host cell's DNA. When the host cell divides, it copies the virus's DNA along with its own. Under certain conditions, the cell is forced to produce new viruses—complete with genes and a protein shell—which can then escape to infect a new cell. Retroviruses sometimes trigger cells to turn cancerous if their genetic material is accidentally inserted in the wrong place in their host's genome. Retroviruses have genetic "on switches" that prompt their host cell to make proteins out of nearby genes. Sometimes their switches turn on host genes that ought to be kept shut off, and cancer can result.

Avian leukosis virus proved to be a very strange retrovirus. At the time, scientists tested for the presence of the virus by screening chicken blood for one of the virus's proteins. Sometimes they would find the avian leukosis virus protein in the blood of chickens that were perfectly healthy and never developed cancer. Stranger still, healthy hens carrying the protein could produce chicks that were also healthy and also carried the protein.

Robin Weiss, a virologist then working at the University of Washington, wondered if the virus had become a permanent, harmless part of the chicken DNA. He and his colleagues treated cells from healthy chickens with mutation-triggering chemicals and radiation to see if they could flush the virus out from its hiding place. Just as they had suspected, the mutant cells started to churn out the avian leukosis virus. In other words, these healthy chickens were not simply infected with avian leukosis virus in some of their cells; the genetic instructions for making the virus were implanted in *all* of their cells, and they passed those instructions down to their descendants.

These hidden viruses were not limited to just one oddball breed of chickens. Weiss and other scientists found avian leukosis virus embedded in many breeds, raising the possibility that the virus was an ancient component of chicken DNA. To see just how long ago avian leukosis viruses infected the ancestors of today's chickens, Weiss and his colleagues traveled to the jungles of Malaysia. There they trapped red jungle fowl, the closest wild relatives of chickens. The red jungle fowl carried the same avian leukosis virus, Weiss found. On later expeditions, he found that other species of jungle fowl lacked the virus.

Out of the research on avian leukosis virus emerged a hypothesis for how it had merged with chickens. Thousands of years ago, the virus plagued the common ancestor of domesticated chickens and red jungle fowl. It invaded cells, made new copies of itself, and infected new birds, leaving tumors in its wake. But in at least one bird, something else happened. Instead of giving the bird cancer, the virus was kept in check by the bird's immune system. As it spread harmlessly through the bird's body, it infected the chicken's sexual organs. When an infected bird mated, its fertilized egg also contained the virus's DNA in its own genes.

As the infected embryo grew and divided, all of its cells also inherited the virus DNA. When the chick emerged from its shell, it was part chicken and part virus. And with the avian leukosis virus now part of its genome, it passed down the virus's DNA to its own offspring. The virus remained a silent passenger from generation to generation for thousands of years. But under certain conditions, the virus could reactivate, create tumors, and spread to other birds.

Scientists recognized that this new virus was in a class of its own. They called it an endogenous retrovirus—endogenous meaning *generated within*. They soon found endogenous retroviruses in other animals. In fact, the viruses lurk in the genomes of just about every major group of vertebrates, from fish to reptiles to mammals. Some of the newly discovered endogenous retroviruses turned out to cause cancer like avian leukosis virus. But many did not. They were crippled with mutations that robbed them of the ability to make new viruses that could escape their host cell. These hobbled viruses could still make new copies of their genes, though, which were reinserted back into their host's genome. And scientists also discovered other endogenous retroviruses that were so riddled with mutations that they could no longer do anything at all. They had become nothing more than baggage in their host's genome.

As far as scientists can tell, no endogenous retroviruses in the human genome are active. But Thierry Heidmann, a researcher at the Gustave Roussy Institute in Villejuif, France, and his colleagues have discovered that they can transform this genetic baggage back into full-blown viruses. Heidmann was studying an endogenous retrovirus when he noticed that different people had slightly different versions. These differences presumably arose after a retrovirus became trapped in the genomes of ancient humans. In their descendants, mutations struck different parts of the virus's DNA.

Heidmann and his colleagues compared the variants of the virus-like sequence. It was as if they found four copies of a play by Shakespeare, each transcribed by a slightly careless clerk. Each clerk might make his own set of mistakes. Each copy might have a different version of the same word—say, *wheregore*, *sherefore*, *whorefore*, *wherefrom*. By comparing all four versions, a historian could figure out that the original word was *wherefore*.

Using this method, Heidmann and his fellow scientists were able to use the mutated versions in living humans to determine the original sequence of the DNA. They then synthesized a piece of DNA with a matching sequence and inserted it into human cells they reared in a culture dish. Some of the cells produced new viruses that could infect other cells. In other words, the original sequence of the DNA had been a living, functioning virus. In 2006, Heidmann named the virus Phoenix, for the mythical bird that rose from its own ashes.

The Phoenix virus probably infected our ancestors within the last million years. But we also carry far older viruses, too. We know this because scientists have found the same viruses lurking in our own genome and in those of other species. Adam Lee, a virologist at Imperial College London, and his colleagues, for example, discovered an endogenous retrovirus called ERV-L in the human genome. They then discovered it in many other species ranging from horses to aardvarks. When Lee and his colleagues drew out an evolutionary tree of the virus, it mirrored the tree of the virus's hosts. It appears that this endogenous retrovirus infected the common ancestor of all mammals with placentas, which lived over 100 million years ago. Today, that virus lingers on, in armadillos and elephants and manatees. And in us.

As an endogenous retrovirus gets trapped in its host, it can still make new copies of its DNA, which get inserted back into its host's genome. Over the millions of years that endogenous retroviruses have been invading our genomes, they've accumulated to a staggering extent. Each of us carries almost a hundred thousand fragments of endogenous retrovirus DNA in our genome, making up about 8 percent of our DNA. To put that figure in perspective, consider that the twenty thousand protein-coding genes in the human genome make up only 1.2 percent of our DNA. Scientists have also cataloged millions of small pieces of DNA that also get copied and inserted back in the human genome. It's possible that many of those pieces descend from endogenous retrovirus. Over millions of years, evolution stripped them down to the bare essentials required for copying DNA. Our genomes, in other words, are awash in viruses.

While most of this viral DNA is useless, our ancestors have commandeered some of it for our own benefit. In fact, without these viruses, none of us today would have been born.

In 1999, Jen-Luc Blond and his colleagues discovered a human endogenous retrovirus they dubbed HERV-W. And they were surprised to discover that one of the genes could still produce a protein. The protein, called syncytin, turned out to have a very precise, very important job to do—not for the virus, but for its human host. It could be found only in the placenta.

The cells in the outer layer of the placenta made syncytin in order to join together, so that molecules could flow between them. Scientists discovered that mice made syncytin too, a discovery that allowed them to run experiments to understand how the protein worked. When they deleted the gene for syncytin, mouse embryos never survived to birth. The viral protein was essential for drawing in nutrients from their mother's bloodstream.

Scientists looked for syncytin in other placental mammals, and they found them. But this surprising protein had one more surprise to deliver: it was actually more than one protein. Different endogenous retroviruses had infected different placental mammal lineages. In some lineages—including our own—two viruses had delivered two different syncytin proteins, the newer one replacing the older.

Thierry Heidmann, who has discovered many of these syncytin proteins, has proposed a scenario to make sense of all these viruses in the placenta. Over 100 million years ago, an ancestral mammal was infected by an endogenous retrovirus. It harnessed the original syncytin protein, and evolved the very first placenta. Over the course of millions of years, that original placental mammal gave rise to many lineages of descendants. And they continued to be infected with endogenous retroviruses. In some cases, the new viruses had syncytin genes as well, which produced superior proteins for the placenta. Different lineages of mammals—rodents, bats, cows, primates—swapped one viral protein for another.

In our most intimate moment, as new human life emerges from old, viruses are essential to our survival. There is no us and them—just a gradually blending and shifting mix of DNA.

THE VIRAL FUTURE

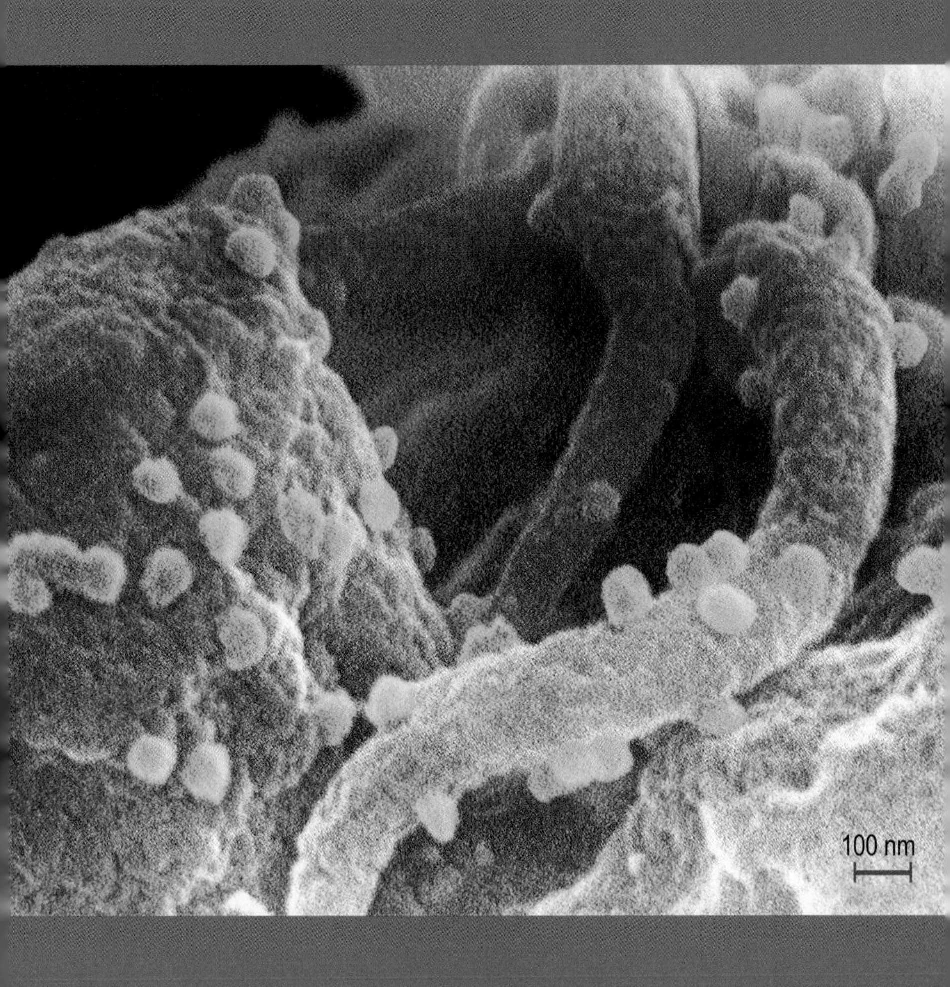

Human immunodeficiency viruses (HIV) on the surface of a CD4 white blood cell

The Young Scourge

Human Immunodeficiency Virus and the Animal Origins of Diseases

Every week, the Centers for Disease Control and Prevention (CDC) publishes a thin newsletter called *Morbidity and Mortality Weekly Report*. The issue that appeared on July 4, 1981, was a typical assortment of the ordinary and the mysterious. Among the mysteries that week was a report from Los Angeles, where doctors had noticed an odd coincidence. Between October 1980 and May 1981, five men were admitted to hospitals around the city with the same rare disease, known as pneumocystis pneumonia.

Pneumocystis pneumonia is caused by a common fungus called *Pneumocystis jiroveci*. The spores of *P. jiroveci* are so abundant that most people inhale it at some point during their childhood. Their immune system quickly kills off the fungus and produces antibodies that ward off any future infection. But in people with weak immune systems, *P. jiroveci* runs rampant. The lungs fill with fluid and become badly scarred. Pneumocystis pneumonia leaves its victims struggling to inhale enough oxygen to stay alive. The five Los Angeles patients did not fit the typical profile of a pneumocystis pneumonia victim. They were young men who had been in perfect health before they came down with pneumonia. Commenting on the report, the editors of *Morbidity and Mortality Weekly Report* speculated that the puzzling symptoms of the five men "suggest the possibility of a cellular-immune dysfunction."

Little did they know that they were publishing the first observations of what would become the greatest epidemic in modern history. The five Los Angeles men did indeed have a cellular-immune dysfunction, caused by a virus that would later be dubbed human immunodeficiency virus (HIV). Researchers would discover that HIV had been secretly infecting people for well over fifty years. After its discovery in the 1980s, the virus went on to infect over sixty million people. It has killed nearly half of them.

HIV's death toll is all the more terrifying because it's actually not all that easy to catch. You can't get HIV if an infected person sneezes near you or shakes your hand. HIV has to be spread through certain bodily fluids, such as blood and semen. Unprotected sex can transmit the virus. Contaminated blood supplies can infect people through transfusions. Infected mothers can pass HIV to their unborn children. Some heroin addicts acquired the virus from infected needles they shared with others.

Once HIV gets into a person's body, it boldly attacks the immune system itself. It grabs onto certain kinds of immune cells known as CD4 T cells and fuses their membranes like a pair of colliding soap bubbles. Like other retroviruses, it inserts its genetic material into the cell's own genome. Its genes and proteins manipulate and then take over the cell, causing it to make new copies of HIV, which escape and can infect other cells.

At first, the population of HIV in a person's body explodes rapidly. Once the immune system recognizes infected cells it starts to kill them, driving the virus's population down. To the infected person, the battle feels like a mild flu. The immune system manages to exterminate most of the HIV, but a small fraction of the viruses manages to survive by lying low. The CD4 T cells in which they hide continue to grow and divide. From time to time, an infected CD4 T cell wakes up and fires a blast of viruses that infect new cells. The immune system attacks these new waves, but over time it becomes exhausted and collapses.

It may take only a year for an immune system to fail, or more than twenty. But no matter how long it takes, the outcome is the same: people can no longer defend themselves against diseases that would never be able to harm a person with a healthy immune system. In the early 1980s, a wave of HIV-infected people began to come to hospitals with strange diseases like pneumocystis pneumonia.

Doctors discovered the effects of HIV before they discovered the virus itself. They dubbed the disease acquired immunodeficiency syndrome, or AIDS. In 1983, two years after the first AIDS patients came to light, French scientists isolated HIV from a patient with AIDS for the first time. More research firmly established HIV as the cause of AIDS. Meanwhile, doctors were discovering more cases of AIDS, both in the United States and abroad. Other great scourges, such as malaria and tuberculosis, are ancient enemies, which have been killing people for thousands of years. Yet HIV went from utter obscurity in 1980 to a global scourge in a matter of a few years. Here was an epidemiological mystery.

It would take scientists three decades to work out the origins of HIV in their broad outlines. The first clues came from sick monkeys. At primate research centers around the United States, pathologists noticed that a number of animals were developing an AIDS-like condition. They wondered if the monkeys were infected with an HIV-like virus. In 1985, scientists at the New England Regional Primate Research Centers mixed HIV antibodies into the serum of sick monkeys and found that the molecules stuck to a virus that was new to science. It came to be known as simian immunodeficiency virus, or SIV for short. Further studies revealed that dozens of other SIV strains infected other species of monkeys and apes. Their discovery raised the possibility that HIV evolved from a strain of SIV—although it wasn't clear at first which strain was the ancestor.

In 1991, Preston Marx of New York University and his colleagues found an extremely HIV-like virus in a West African monkey known as the sooty mangabey. But this discovery didn't resolve the question of HIV's origin so much as deepen it. In the 1980s, scientists had found that they could neatly sort HIV into two genetically distinct clusters. They named these two types HIV-1 and HIV-2. HIV-1 was common around the world, while HIV-2 could be found only in parts of West Africa, where it was much less aggressive. The strain of SIV Marx found in sooty mangabeys was closely related to HIV-2—more closely related, in fact, than HIV-1.

Marx's discovery hinted that HIV did not have a single origin. Instead, HIV-2 had evolved independently from sooty mangabey SIV, while HIV-1 had evolved from a yet-to-be-discovered strain. As scientists studied sooty mangabey SIV further, the picture grew even more complex. They discovered new SIV strains that were more closely related to particular lineages of HIV-2 than other HIV-2 lineages. These findings revealed that sooty mangabey SIV had repeatedly leaped into humans to produce HIV-2—nine times, all told.

No one witnessed those nine leaps, but we can be fairly certain how they happened. In West Africa, many people keep sooty mangabeys as pets. It's also common for hunters to kill the monkeys and sell their meat. The virus could move from sooty mangabeys to people whenever their blood made contact—when a monkey bit a hunter, for example, or when a butcher prepared its meat. The SIV could then infect host cells, replicate, and adapt to its new host species.

Deciphering the origins of HIV-1, the major source of infection, took longer. In 1989, Martine Peeters, a virologist at the Institute of Research for Development and the University of Montpellier in France, and her colleagues discovered an HIV-1-like virus in captive chimpanzees in Gabon. To survey the virus in the wild, the researchers headed to forests across equatorial Africa. They didn't try to get blood samples from chimpanzees, since the apes are elusive, strong, and not fond of people with needles. Instead, they collected feces that the apes deposited underneath the trees where they slept. Back in their labs, the scientists search the stool for viruses. They isolated SIV strains from across the entire range of chimpanzees, from Cameroon in the east to Tanzania in the west. But they found close relatives to HIV-1 only in chimpanzees that lived in southern Cameroon. In 1999, when the scientists drew the evolutionary tree of the viruses, they found that HIV-1's history was similar to that of HIV-2, with several origins instead of one.

HIV researchers had sorted the world's HIV-1 viruses into four groups. Group M—which is short for "main"—is responsible for 90 percent of HIV-1 cases. The other groups, known as N, O, and P, are rarer. Peeters and her colleagues discovered that one chimpanzee SIV strain from Cameroon was closely related to Group M. Another chimpanzee virus matched Group N. Their discovery left Group O and Group P unaccounted for.

As Peeters and her colleagues tramped through forests searching for chimpanzee stool, they also kept an eye out for other primates. In some of the forests they visited, gorillas also made their nests. The scientists scooped up gorilla stool and brought it back to their lab to study as well. And in 2006, they announced that Cameroon gorillas had SIV, too.

The gorilla SIV, they found, had originated from chimpanzees. To better understand how gorilla SIV was related to viruses in chimpanzees and humans, Peeters and her colleagues set out to systematically gather gorilla stool from across their range as well. Searching through more than 3,000 samples of gorilla feces, they failed to find any gorilla SIV outside of Cameroon. But inside Cameroon, the story was different. They found more gorilla SIV. In 2015, Peeters and her colleagues announced that two of these new lineages of gorilla SIV were the ancestors of HIV-1 O and P, the last two groups of HIV-1 to be pinned down.

Group P is the rarest form of HIV ever found, isolated only from two people from Cameroon. As an evolutionary experiment, HIV-1 Group P seems to have been a failure, producing a virus poorly adapted to infecting humans. Group O, on the other hand, made a more impressive jump. It has infected an estimated 100,000 people in Cameroon. When scientists have studied the biology of the virus, they've found that it's just as good at multiplying inside humans as HIV-1 Group M. With the discovery of gorilla SIV, the overall shape of HIV's evolutionary tree is now clear. SIV has leaped thirteen times into our species—nine times as HIV-2 strains, and four times as HIV-1.

But when did these leaps happen? To pin down their timing, some scientists have looked back at patients who had died mysteriously before HIV was even discovered. In 1988, for example, researchers found that a Norwegian sailor named Arvid Noe, who died in 1976, had been infected with HIV-1 Group O. In 1998, David Ho and his colleagues at Rockefeller University isolated HIV-1 Group M from a blood sample taken from a patient in Kinshasa in 1959. In 2008, Michael Worobey and his colleagues at the University of Arizona discovered another sample of HIV-1 Group M in a second tissue sample from another pathology collection in Kinshasa, dating back to 1960.

To reach further back, scientists have extracted the history encoded in HIV's genes. As the virus replicates, it accumulates mutations at a clock-like rate, piling up like sand in an hourglass. By measuring the height of this genetic sandpile, scientists can estimate how much time had passed. Using this method, scientists have found that both HIV-1 Group M and HIV-1 Group O originated in the early 1900s. (There isn't enough data to estimate the dates for the other lineages of HIV.)

All this evidence points to a compelling hypothesis for how HIV-1 got its start—or rather, its starts. HIV-1–like viruses had been circulating among populations of chimpanzees throughout Africa. At some point, a gorilla contracted SIV from a chimpanzee—perhaps during a bloody fight over a tree full of ripe figs in Cameroon. For centuries, hunters in Cameroon would kill both chimpanzees and gorillas for meat. And from time to time they would become infected with SIV from the apes. But these hunters, living in relative isolation before the twentieth century, were a dead end for the viruses. Some people recovered from SIV infections because their immune systems could vanquish the poorly adapted hosts.

In other cases, the viruses were extinguished with the death of their host, unable to reach a new one.

Africa started undergoing dramatic changes in the early 1900s that gave SIV new opportunities to spread into humans. Commerce along the rivers allowed people to move from villages to towns, bringing their viruses with them. The colonial settlements in central Africa began to expand to cities of ten thousand people or more, giving the virus more opportunities to spread from host to host. One strain of chimpanzee SIV infiltrated this growing population and evolved into HIV-1 Group M. And from a gorilla, another strain adapted to humanity: HIV-1 Group O.

As these two viruses replicated in their human hosts, they mutated, and some of those mutations made it easier for the viruses to proliferate. To become successful human viruses, each strain had to overcome powerful human immune defenses. When HIV replicates, the new viruses need to escape their host cell to continue proliferating. Our cells make a protein called tetherin that grabs hold of the viruses and keep them anchored to the cell where they formed. The two strains of HIV-1 evolved different strategies for evading tetherin's grasp.

For their first few decades, HIV-1 Group M and Group O both grew slowly in Cameroon. Group O, the gorilla virus, never managed to escape. But Group M got a lucky break. Somehow the virus traveled in the mid-1900s to Kinshasa (known then as Leopoldville). In the dense city slums, the virus spread quickly. Infected people traveled from the city along rivers and train lines to the other big cities of central Africa, such as Brazzaville, Lubumbashi, and Kisangani. By 1960 HIV-1 Group M straddled Africa's waistline.

In the next few years, HIV-1 Group M spread to Haiti, as Haitians who had been working in the Congo returned to their homeland after the country became independent from Belgium. Later, Haitian immigrants or American tourists may have brought HIV to the United States by the 1970s. That's about four decades after the virus became established in humans, and about one decade before five men in Los Angeles became sick with a strange form of pneumonia.

By the time scientists recognized HIV in 1983, in other words, the virus had already become a hidden global catastrophe. And by the time scientists began trying to fight the virus, HIV already had a huge head start. The annual death toll climbed through the 1980s and 1990s. Some scientists predicted the virus could be quickly stopped with a vaccine, but a series of failed experiments dashed their hopes.

It took years of hard work to stem the tide of HIV. Public health workers found that they could dial down the transmission of the virus with social policies, such as controlling the use of needles and distributing condoms. Later, the invention of powerful anti-HIV drugs helped the fight enormously. Today, millions of people take a cocktail of drugs that interfere with the ability of HIV to infect immune cells and use them to replicate. In affluent countries such as the United States, these drug therapies have allowed some people to enjoy a relatively healthy life. As governments and private organizations deliver those drugs to poorer countries, victims of HIV are living longer there as well. In 2005, the annual death rate from HIV peaked at 2.5 million people a year. Since then, it's slowly declined. By 2013, it was down to 1.5 million people.

In theory, the world could drive that number to zero. An HIV vaccine still remains the best hope to reach that goal, and recent research has renewed hope that an effective formula may arrive soon. Increasing the use of anti-HIV drugs could also help, by lowering the number of viruses in people who are already infected. But many researchers are also exploring HIV's biology and history in the hope of finding some feature of the virus that could become an Achilles's heel. We now understand HIV so well that we can see the molecular steps by which it adapted to our species a century ago. It may be possible to undermine those adaptations. The future of fighting HIV, in other words, may perhaps lie in its past.

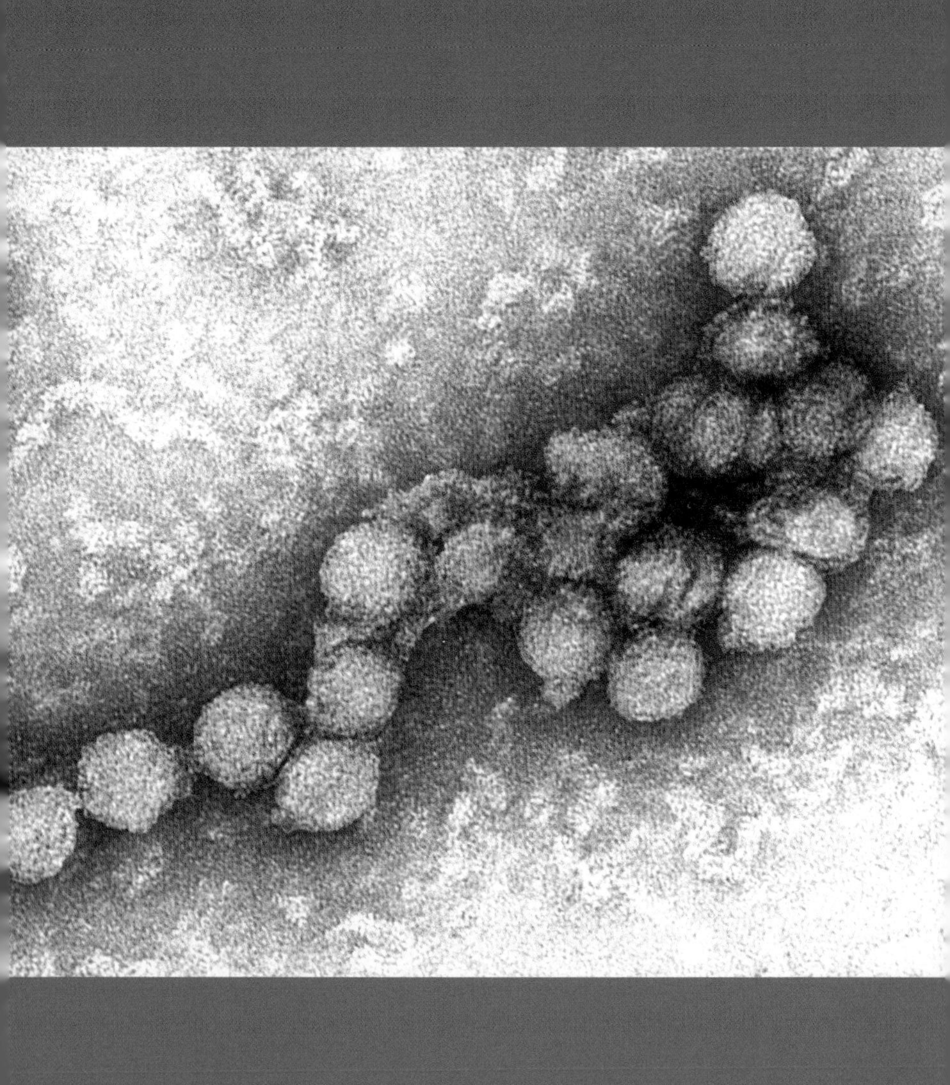

West Nile viruses in suspension

Becoming an American

The Globalization of West Nile Virus

In the summer of 1999, the crows started to die.

Tracey McNamara noticed the dead crows lying around the Bronx Zoo, where she worked as the chief pathologist. The sight of the birds made her worried. McNamara feared that some new virus was spreading from bird to bird across New York City. If the crows were dying, the zoo's birds might start to die, too.

Over Labor Day weekend, her worst fears were realized. Three flamingoes suddenly died. So did a pheasant, a bald eagle, and a cormorant. Zoo workers brought McNamara the dead birds, and she examined them to figure out what had killed them. They showed clear signs of some kind of infection that caused their brains to bleed. But McNamara could not figure out what pathogen was responsible, so she sent tissue samples to government laboratories. The government scientists ran test after test for the various pathogens that might be responsible. For weeks, the tests kept coming up negative.

Meanwhile, doctors in Queens were seeing a worrying number of cases of encephalitis—an inflammation of the brain. The entire city of New York normally sees only nine cases a year, but in August 1999, doctors in Queens found eight cases in one weekend. As the summer waned, more cases came to light. Some patients suffered fevers so dire that they became paralyzed, and by September nine had died. Initial tests pointed to a viral disease called Saint Louis encephalitis, but later tests failed to match the results.

As doctors struggled to make sense of the human outbreak, McNamara was finally getting the answer to her own mystery. The National Veterinary Services Laboratory in Iowa managed to grow viruses from the bird tissue samples she had sent them from the zoo. They bore a resemblance to the Saint Louis encephalitis virus. McNamara wondered now if both humans and birds were succumbing to the same pathogen. She convinced the Centers for Disease Control and Prevention to analyze the genetic material in the viruses. On September 22, the CDC researchers were stunned to find that the birds had not died of Saint Louis encephalitis. Instead, the culprit was a pathogen called West Nile virus, which infects birds as well as people in parts of Asia, Europe, and Africa. No one had imagined that the Bronx Zoo birds were dying of West Nile virus, because it had never been seen in a bird in the Western Hemisphere before.

Public health workers puzzling over the human cases of encephalitis decided it was time to broaden their search as well. Two teams—one at the CDC and another led by Ian Lipkin, who was then at the University of California, Irvine—isolated the genetic material from the human viruses. It was the same virus that was killing birds: West Nile. And once again, it took researchers by surprise. No human in North or South America had ever suffered from it before.

The United States is home to many viruses that make people sick. Some are old and some are new. When the first humans made their way into the Western Hemisphere some fifteen thousand years ago, they brought a number of viruses with them. Human papillomavirus, for example, retains traces of its ancient emigration. The strains of the virus found in Native Americans are more closely related to each other than they are to HPV strains in other parts of the world. Their closest relative outside of the New World are strains of HPV found in Asia, just as Native Americans are most closely related to Asians.

When Europeans arrived in the New World, they brought a second wave of viruses with them. New diseases such as influenza and smallpox wiped out millions of Native Americans. In later centuries, still more viruses arrived. HIV came to the United States in the 1970s, and at the end of the twentieth century, West Nile virus became one of America's newest immigrants.

It had been only six decades since West Nile virus was discovered anywhere on the planet. In 1937, a woman in the West Nile district of Uganda came to a hospital with a mysterious fever, and her doctors isolated a new virus from her blood. Over the next few decades, scientists found the same virus in many patients in the Near East, Asia, and Australia. But they also discovered that West Nile virus did not depend on humans for its survival. Researchers detected the virus in many species in birds, where it could multiply to far higher numbers.

At first it was not clear how the virus could move from human to human, from bird to bird, or from bird to human. That mystery was solved when scientists found the virus in a very different kind of animal: mosquitoes. When a virus-bearing mosquito bites a bird, it sticks its syringe-like mouth into the animal's skin. As the mosquito drinks, it squirts saliva into the wound. Along with the saliva comes the West Nile virus.

The virus first invades cells in the bird's skin, including immune system cells that are supposed to defend animals from diseases. Virus-laden immune cells crawl into the lymph nodes, where they release their passengers, leading to the infection of more immune cells. From the lymph nodes, infected immune cells spread into the bloodstream and organs such as the spleen and kidneys. It takes just a few days for the viruses in a mosquito bite to multiply into billions inside a bird. Despite their huge numbers, West Nile viruses cannot escape a bird on their own. They need a way out—what scientists call a vector. A mosquito must bite the infected bird, drawing up some of its virus-laden blood. Once in the mosquito, the viruses invade the cells of its midgut. From there they can be carried to the insect's salivary glands, where the viruses are ready to be injected into a new bird.

Vector-borne viruses like West Nile virus require a special versatility to complete their life cycle. Mosquitoes and birds are profoundly different kinds of hosts, with different body temperatures, different immune systems, and different anatomies. West Nile virus has to be able to thrive in both environments to complete its life cycle. Vector-borne viruses also pose special challenges to doctors and public health workers who want to stop their spread. They don't require people to be in close contact to spread from host to host. Mosquitoes, in effect, give the viruses wings.

Studies on the genes of West Nile virus suggest that it first evolved in Africa. As birds migrated from Africa to other continents in the Old World, they spread the virus to new bird species. Along the way, West Nile virus infected humans. In Eastern Europe, epidemics broke out, producing some cases of encephalitis. In a 1996 epidemic in Romania, ninety thousand people came down with West Nile, leading to seventeen deaths. These new epidemics, first in Europe and later in the West, may have been the result of the virus infecting people whose populations had not experienced it before. In Africa, by contrast, people may be immunized against West Nile virus after being infected while they're young.

It is striking that the New World has been spared West Nile virus for so long. The flow of people across the Atlantic and Pacific was not enough to carry the virus to the Americas. Scientists cannot say exactly how West Nile virus finally landed in New York in 1999, but they have a few clues. The New World strain of West Nile virus is most closely related to viruses that caused an outbreak in birds in Israel in 1998. It's possible that pet smugglers brought infected birds from the Near East to New York.

On its own, a single infected bird could not have triggered a nationwide epidemic. The viruses needed a new vector to spread. It just so happens that West Nile viruses can survive inside 62 species of mosquitoes that live in the United States. The birds of America turned out to be good hosts as well. All told, 150 American bird species have been found to carry West Nile virus. A few species, such as robins, blue jays, and house finches, turned out to be particularly good incubators.

Moving from bird to mosquito to bird, West Nile virus spread across the entire United States in just 4 years. And along the way, millions of people were infected. Only about 25 percent of people who acquire the virus develop a fever. Between 1999 and 2013, scientists estimate that over 780,000 people became sick with West Nile virus. Of those, 16,196 went on to develop encephalitis, and there were 1,549 reported deaths.

Once West Nile virus arrived in the United States, it settled into a regular cycle, a cycle set by the natural history of birds and mosquitoes. In the spring, birds produce new generations of chicks that are helpless targets for virus-carrying mosquitoes. The percentage of infected birds goes up through the summer, and many mosquitoes get infected by feeding on them. Those mosquitoes can then bite people who are spending more time outdoors in the warm months of the year, giving them West Nile.

When the temperature drops in the fall, mosquitoes die in much of the United States, and the viruses can no longer spread. No one knows for sure how the virus survives the winters without an insect host. It's possible that they survive in low levels among mosquitoes in the south, where the winters aren't so harsh. It's also possible that mosquitoes infect their own eggs with West Nile virus. When infected eggs hatch the next spring, the new generation is ready to start infecting birds all over again.

West Nile virus has fit so successfully into the ecology of the United States that it's probably going to be impossible to eradicate. Even if doctors could give out a vaccine that would prevent any human being from becoming a host for the virus, it could still thrive in birds. And the sad truth is that there is no approved vaccine for West Nile virus, and there may never be. That's because the virus infects a huge number people but only causes a relatively modest cases of encephalitis, and even fewer deaths. Vaccinating a large portion of the United States population would be tremendously expensive, far greater than the cost of hospitalizing the people who get sick from the virus.

The story of West Nile Virus is now replaying itself with the arrival of a new pathogen in the New World. In December 2013, a new mosquito-borne virus turned up in the Caribbean, called chikungunya. Before then, the virus had been recorded only in the Old World, where it caused painful, arthritis-like symptoms. Nobody can say how chikungunya virus arrived in the New World—whether it was a virus-infected traveler or a mosquito released from an airplane. The one clue scientists do have is the genetic material of the viruses. The Caribbean strain of chikungunya is nearly identical to a strain that has circulated in China and the Philippines. Somehow the virus leaped across the planet.

Once it made the leap, it exploded. In its first year alone in the New World, chikungunya caused over a million infections. By the end of 2014, it had spread to many islands in the Caribbean, taxing their health care systems. But it had not yet established itself in North or South America. Public health experts take little comfort from that fact. Mosquitoes that can carry the virus thrive on the continents, and if they became infected, they could easily spread it to millions of people. And with heavy traffic by plane and boat from the Caribbean to the mainland, it's just a matter of time before some people deliver enough of the viruses home to start a new epidemic.

The future looks rosy for West Nile virus and other mosquito-borne viruses that follow it to the New World. That's because the future is going to be warm and wet. Carbon dioxide and other heat-trapping gases are raising the average temperature in the United States. Climate scientists project that the temperature will continue to rise much higher in decades to come, with some regions expecting more humid, stormy conditions. Jonathan Soverow of Beth Israel Deaconess Medical Center and his colleagues examined sixteen thousand cases of West Nile virus that occurred between 2001 and 2005, noting the weather at the time of each outbreak.

They found that epidemics tended to occur when there was heavy rainfall, high humidity, and warm temperatures. Warm, rainy, muggy weather makes mosquitoes reproduce faster and makes their breeding season longer. It also speeds up the growth of the viruses inside the mosquitoes. Now that West Nile virus has made a new home here, we're making that home more comfortable.

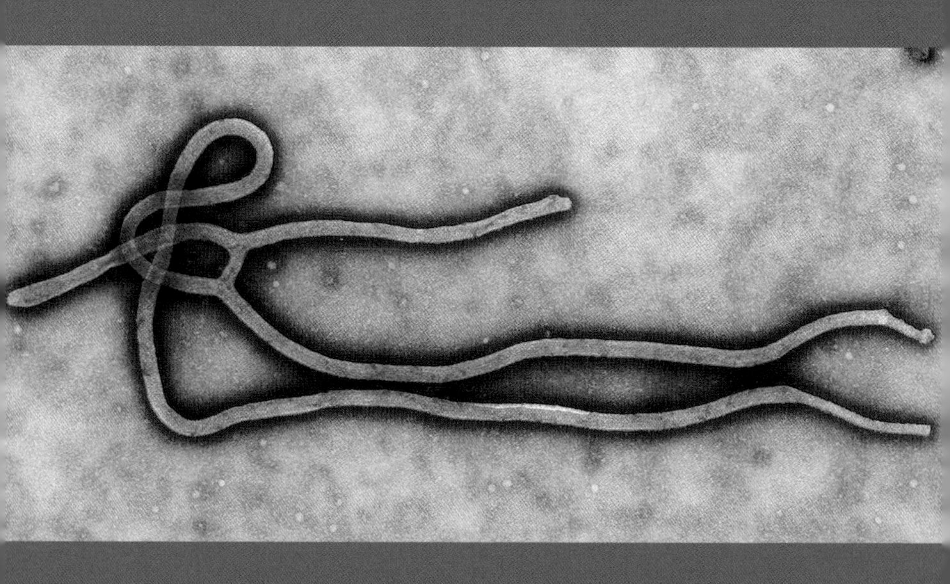

An Ebola virus in suspension

Predicting the Next Plague

Ebola Virus and the Many Others Like It

On December 2, 2013, in the village of Meliandou in southeastern Guinea, a two-year-old boy fell ill. Emile Ouamouno developed a fever, and then started vomited violently and suffering from explosive, bloody diarrhea. No one in Meliandou had seen anything like his sickness before. His family tried to care for him as best they could, but, on December 6, he died. A virus had killed the boy, and by the time he died, the virus's descendants had spread to other members of his family. Emile's four-year-old sister Philomène soon developed the same horrific symptoms and died. So did his mother and grandmother. They probably became infected through the most cruel route imaginable—by caring for the dying boy.

If the virus had stopped there, this family tragedy might have gone unnoticed beyond Meliandou. After all, people die from many viruses and other pathogens in Guinea at a high rate. But this virus was different. It was exquisitely lethal, killing about seventy percent of the people it infected. The Ouamouno family infected a nurse and the village midwife, who also became ill. The midwife was taken back to her home village, Dandou Pombo, to be cared for by her family, where the virus spread further. Meanwhile, some of the people who had attended the funeral of Emile Ouamouno's grandmother returned to their own villages, only to become sick as well.

And soon the outbreak became international. Meliandou is located near the borders of Guinea, Sierra Leone, and Liberia, and people regularly travel from country to country to do business or to see family. Soon Sierra Leone and Liberia had their own outbreaks. But because the virus was still spreading among remote villages nestled in rain forests, no one in the outside world took it as seriously as they should have. It took until March 2014 for medical authorities in Guinea to finally announce the cause of the outbreak: a virus called Ebola.

Some viruses are old enemies of our species. Rhinoviruses gave ancient Egyptians colds, and endogenous retroviruses invaded the genomes of our primate ancestors tens of millions of years ago. Other viruses are younger. HIV, for example, became a human virus about a century ago. And other viruses are only just starting to patter down on our species, triggering outbreaks and raising fears of new global plagues. Of all these newly discovered viruses, none inspires more fear for the future than Ebola.

Ebola made its terrifying debut in 1976. In a remote region of Zaire, people began to develop fevers and vomit. Some victims bled uncontrollably out of every orifice, even from their eyes. A doctor treating a dying nun put samples of her blood in a thermos, which he had delivered to Kinshasa, and then flown to Belgium. There, a young virologist named Peter Piot examined infected cells with an electron microscope and saw swarms of snake-shaped viruses.

At the time, virologists only knew of one other snake-shaped virus: a dangerous pathogen called Marburg virus. It gained its name from the Germany city of Marburg, where lab workers fell ill with a hemorrhagic fever after handling monkeys imported from Uganda. But Piot could tell that what he was looking at was not Marburg virus, but a new relative of it. Piot and his colleagues were so alarmed by the prospect of a deadly new disease that they made their way to Zaire, and eventually to the village of Yambuku, to investigate the outbreak. They found a guesthouse where nuns and priests had holed themselves up, with a rope blocking any visitors. A sign that hung from the rope read, "Please stop. Anybody who crosses here may die."

Piot and his team conducted an epidemiological survey, determining who was infected, and when they had become sick. Soon they had uncovered the paths that the virus, still unnamed, was taking from person to person. It showed no sign of floating through the air like influenza or measles viruses. Instead, it traveled in the fluids of victims. A local hospital that reused syringes spread the virus to many patients. People who cared for the sick and bathed the bodies of the dead also became sick.

As lethal as Ebola could be, the route of its spread was relatively easy to interrupt. Piot and his colleagues closed the hospital and quarantined anyone showing the symptoms of the disease. After three months the outbreak ended, with 318 deaths. Without Piot's intervention, it might have become far worse. His last act on his trip was to give the virus its name. He didn't want to saddle Yambuku with the notoriety, and so he looked to a nearby river, the Ebola.

In the same year, Ebola emerged again in Sudan, claiming 284 lives. It reappeared in Sudan three years later, killing 34. Then it disappeared for fifteen years, striking again in Gabon in 1994, where it killed 52 people. With each flare-up, Piot's successors learned more about how to stop Ebola outbreaks, simply tracking patients and isolating them to stop new infections. They had no vaccine to offer, no antiviral drugs to cure the infected.

Many viruses can cause sudden outbreaks when the conditions are right. But after an outbreak, a virus like measles or chickenpox doesn't vanish from our species. It just subsides, still circulating at a low frequency. Ebola proved to be different. When an outbreak ended, the Ebola virus seemed to disappear. Years would pass before it returned in a deadly blaze.

A number of virologists began searching for where Ebola went during those missing years. They found that gorillas and chimpanzees were getting infected, too, and were dying at high rates from Ebola. They also found antibodies to Ebola in bats, which seemed to tolerate the virus. It's possible that Ebola normally circulates harmlessly from bat to bat, spilling over into humans from time to time.

One thing about Ebola is clear: it is an old virus, despite being new to us. Evolutionary biologists have discovered Ebola-like genes embedded in the genomes of hamsters and voles. Like endogenous retroviruses, these ancestral relatives of Ebola infected their rodent hosts and accidentally left behind a genetic legacy. Hamsters and voles shared a common ancestor over 16 million years ago, which means that Ebola diverged from its closest relative—Marburg virus—at least that long ago.

For millions of years, in other words, viruses on Ebola's lineage have circulated among mammal hosts. In some species, they spread harmlessly, sometimes spilling over into other species, where their biology turned them lethal. We humans have become Ebola's latest spillover host. It's possible that people pick up Ebola virus when they eat infected meat, or perhaps fruit that's been contaminated by bat saliva. However Ebola gets into our bodies, it promptly invades our immune cells and quickly causes massive amounts of inflammation. The virus makes people fatally leaky. They release huge quantities of diarrhea, vomit, and sometimes blood.

Once Ebola gets into one person, its fate depends on the actions of the people around its first human victim. If they come into contact with an infected person, they can acquire Ebola and spread it to others. For the first 37 years of Ebola's recorded history in humans, the virus burned itself out within months of a new outbreak, as its hosts either died or recovered.

But in those 37 years, the human environment changed. Africa had a population of 221 million in 1950. Today, its population stands at just over a billion. It used to be difficult for Ebola to spread to more than a few neighboring villages. Today, roads cut through more and more rain forest, enabling people to move to towns and cities, where Ebola can find many more hosts to infect. But in countries like Guinea, Liberia, and Sierra Leone, this rapid change has not been accompanied by good public health. Years of civil war and grinding poverty have left these countries with few hospitals or doctors. Once Ebola victims began arriving in hospitals, many health care workers began to die, robbing the countries of the expertise required to stop the outbreak.

No one knows how Emile Ouamouno came down with Ebola, but the virus that spread from him to his family went on to create the biggest Ebola outbreak in history, killing far more people than have died in all the Ebola outbreaks that came before. It spread to the capitals of the three plagued countries—Conakry, Freetown, and Monrovia—and within a few months, hospitals were so crammed with Ebola victims that people began being turned away and sent home to die. International health organizations, distracted by other viruses such as influenza and polio, failed to recognize the true scope of the outbreak and failed to bring help to West Africa that might have slowed it down. The West African governments tried quarantining entire neighborhoods of cities and rural regions, without much to show for the disruptions. The death count climbed, passing a thousand, then two, and onward.

Ebola began to sneak aboard airplanes for the first time. One flight took an infected diplomat to Nigeria, where the virus infected his doctor and others. Another plane took an infected nurse back home to Spain. Two planes brought hidden cases of Ebola to the United States—one to Houston, and one to New York.

Outside of Africa, people knew little about Ebola, and what they did know came mainly from frightening accounts in books like Richard Preston's *The Hot Zone*, or the mostly fictional treatments in movies like *Outbreak*. Fear swept the United States, where a poll conducted in October 2014 revealed that two-thirds of Americans were concerned that an Ebola outbreak would hit the country, and 43 percent worried that they were personally at risk of contracting the disease. Rumors spread that Ebola could travel through the air. On October 23, news broke that a New York doctor who had returned from treating Ebola patients from Guinea had tested positive for the disease. Before his symptoms emerged, he had visited a bowling alley. Worried readers asked the *New York Times* if a bowling ball could spread the virus. Journalist Donald McNeil Jr. quickly replied: "If someone left blood, vomit or feces on a bowling ball, and the next person to touch it did not even notice, and then put his fingers into his eyes, nose or mouth, it might be possible." It was a polite way of saying no.

Despite all the panic, Ebola did not launch massive outbreaks in the United States. Nor did it do so elsewhere. Nigeria suffered 20 cases and 8 deaths before stopping the virus, using the same public health measures Piot and others had developed in earlier outbreaks. Senegal recorded a single case, without any deaths at all. Mali stopped an outbreak of its own. These countries succeeded in their fight against Ebola because they had the luxury of warning. Meanwhile Liberia, Guinea, and Sierra Leone continued to suffer the consequences of the outbreak's hidden beginnings. In those three countries, the fire had grown too big to be easily put out.

Epidemiologists looked anxiously at the skyrocketing case numbers. They projected forward, trying to gauge how bad the outbreak could become. In September 2014, the Centers for Disease Control warned that, without additional intervention, as many as 1.4 million cases of Ebola by January 2015.

Fortunately, a number of countries, including the United States, China, and Cuba, began sending doctors and supplies. New Ebola hospitals were constructed. Public health workers encouraged people to bury Ebola victims safely, so that they would not die as well. The efforts were modest and late, but by late November, some hope glimmered: the rate of new cases began dropping in Liberia and Guinea. In the new year, the outbreak continued to shrink dramatically.

When this Ebola outbreak ends, it won't be the last. In September 2014, a team of ecologists and epidemiologists at the University of Oxford developed a detailed map of the places where Ebola would be most likely to spill over in the future. They took into account the ranges of animal species that can harbor the virus, and the growing human settlements that overlap them. The potential spillover zone forms a thick belt across central Africa, as well as isolated islands of risk in Tanzania, Mozambique, and even Madagascar. All told, 22 million people live in the Ebola-risk zone. While the chances of any one of those 22 million people acquiring Ebola from an animal is very low, the danger posed by such an event is enormous. And as Africa's population grows, that danger will grow too.

And that's just Ebola. Since Ebola was first discovered in 1976, other viruses have made debuts of their own, emerging thousands of miles apart in profoundly different circumstances. In November 2002, for example, a Chinese farmer came to a hospital suffering from a high fever and died soon afterward. Other people from the same region of China began to develop the disease as well, but it didn't reach the world's attention until an American businessman flying back from China developed a fever on a flight to Singapore. The flight stopped in Hanoi, where the businessman died. Soon, people were falling ill in countries around the world, although most of the cases turned up in China and Hong Kong. The disease killed 10 percent of its victims, usually in a matter of days. The outbreak was new to medicine, which meant that it needed a new name. Doctors called it severe acute respiratory syndrome, or SARS.

Scientists began searching samples from SARS victims for a cause of the disease. Malik Peiris of the University of Hong Kong led the team of researchers who found it. In a study of fifty patients with SARS, they discovered a virus growing in two of them. The virus belonged to a group called coronaviruses, which includes species that can cause colds and the stomach flu. Peiris and his colleagues sequenced the genetic material in the new virus and then searched for matching genes in the other patients. They found a match in forty-five of them.

On the basis of their experience with viruses such as HIV and Ebola, scientists suspected that the SARS virus had evolved from a virus that infects animals. They began to analyze viruses in animals with which people in China have regular contact. As they discovered new viruses, they added their branches to the SARS evolutionary tree. In a matter of months, scientists had reconstructed the history of SARS.

The virus likely started in Chinese bats. A lineage of the viruses then began to spill over into a catlike mammal called a civet. Civets are a common sight in Chinese animal markets, and it's likely that humans became spillover hosts as well. The virus turned out to have the right biology for spreading from people to people—and unlike Ebola, it could spread on fine aerosol droplets.

Fortunately, the same public health measures that stopped early Ebola epidemics also managed to stop SARS, even after it spread beyond Asia. It caused about eight thousand cases and nine hundred deaths before it disappeared. Compared to an ordinary year of flu infections—which can kill over 250,000 people—the emergence of SARS was a dodged bullet.

A decade later, another coronavirus appeared in Saudi Arabia. In 2012, doctors at Saudi hospitals noticed that some of their patients were becoming ill with a respiratory disease they couldn't identify. Nearly a third of the patients died. The disease came to be known as MERS, short for Middle Eastern Respiratory Syndrome, and soon after, virologists isolated the virus that caused it. Once scientists could read the genes of the MERS virus, they could search for similar ones in other species. They soon found related viruses in bats in Africa.

How African bats could have caused a Middle Eastern epidemic was a question without an obvious answer. But an important new clue emerged when scientists examined the mammals that many people in the Middle East depend on for their survival: camels. They began to find camels rife with MERS viruses, which oozed from their noses in drops of mucous. One possible explanation for the origin of MERS is that bats passed on the virus to camels in North Africa. There's a healthy trade in camels from North Africa to the Middle East. A sick camel may have carried the virus to its new home.

As scientists reconstructed the history of MERS, there was good reason to fear a global outbreak even worse than SARS. Each year, over two million Muslims travel to Saudi Arabia for the annual pilgrimage known as the hajj. It was easy enough to imagine the MERS virus spreading swiftly among the crowds, and then traveling with the pilgrims to their homes around the world. But so far, that hasn't happened. As of February 2015, 1,026 people have been diagnosed with MERS, 376 of whom have died. Almost all of those cases have occurred in Saudi Arabia—especially in hospitals. It may be that MERS can successfully invade only people with weakened immune systems. Unless MERS undergoes some drastic evolution, it may become a dangerous but rare threat of Middle Eastern hospitals.

We'd be better prepared for these emergencies if they didn't always come as such surprises. The next plague may start when yet another virus in some wild animal jumps into our species—a virus we might not yet even know about. To reduce that ignorance, scientists are surveying animals, searching for bits of genetic material from viruses. But because we live on a planet of viruses, that task is enormous. Ian Lipkin and his colleagues at Columbia University trapped 133 rats in New York City and discovered 18 new species of viruses that are closely related to human pathogens. In another study in Bangladesh, they examined a bat called the Indian flying fox and tried to identify every single virus that calls it home. They identified 55 species, 50 of which are new to science.

We can't say which, if any, of these newly discovered viruses will create a great epidemic. But that doesn't mean that we can simply ignore them. Instead, we need to stay vigilant, so that we can block them before they get a chance to make the great leap into our species.

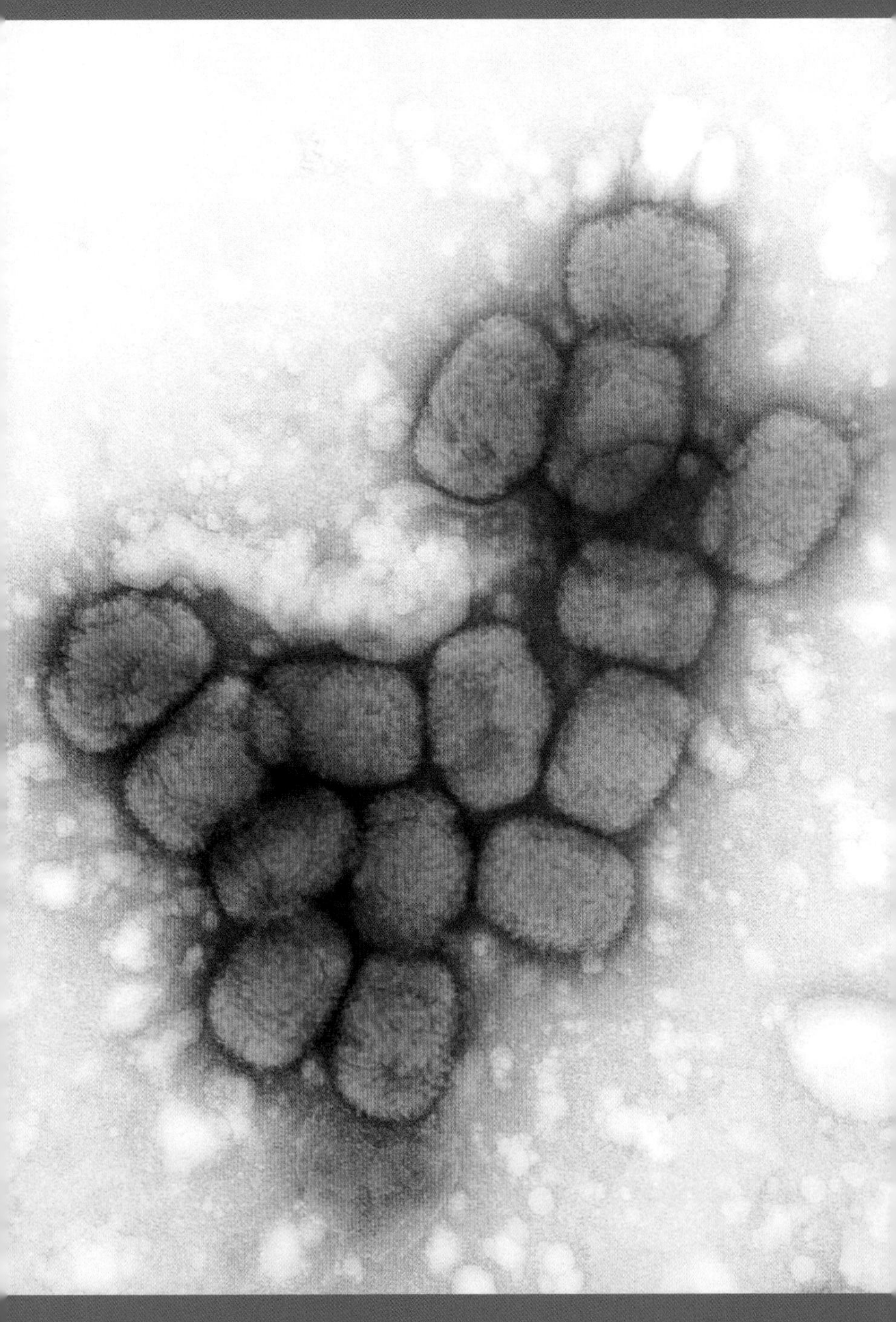

Smallpox viruses in suspension

The Long Goodbye

The Delayed Oblivion of Smallpox

We humans are good at creating new viruses by accident—whether it's a new flu virus concocted on a pig farm, or HIV evolving from the viruses of butchered chimpanzees. What we're not so good at is getting rid of viruses. Despite all the vaccines, antiviral drugs, and public health strategies at our disposal, viruses still manage to escape annihilation. The best we can typically manage is to reduce the harm that viruses cause. HIV infections, for example, have declined in the United States, but fifty thousand Americans still acquire the virus every year. Vaccination programs have eliminated some viruses from some countries, but the viruses can still thrive in other parts of the world. In fact, modern medicine has managed to completely eradicate only a single species of human virus from nature. The distinction goes to the virus that causes smallpox.

But what a virus to wipe out. Over the past three thousand years, smallpox may have killed more people than any other disease on Earth. Ancient physicians were well aware of smallpox, because its symptoms were so clear and distinct. A victim became infected when the virus slipped into the airway. After a week or so, the infection brought chills, a blazing fever, and agonizing aches. The fever ebbed after a few days, but the virus was far from done. Red spots developed inside the mouth, then on the face, and then over the rest of the body. The spots filled with pus and caused stabbing pain. About a third of people who got smallpox eventually died. In the survivors, scabs covered over the pustules, which left behind deep, permanent scars.

Some thirty-five hundred years ago, smallpox left its first recorded trace on humanity: three mummies from ancient Egypt, studded with the scars of pustules. Many other centers of ancient civilization, from China to India to ancient Greece, felt the wrath of the virus. In 430 BC, an epidemic of smallpox swept through Athens, killing a quarter of the Athenian army and a large percentage of the city's population. In the Middle Ages, crusaders returning from the Middle East brought smallpox to Europe. Each time the virus arrived in a new defenseless population, the effects were devastating. In 1241 smallpox first came to Iceland, where it promptly killed twenty thousand of the island's seventy thousand inhabitants. Smallpox became well established in the Old World as cities grew, providing the virus with an easy path from one host to the next. Between 1400 and 1800, smallpox killed an estimated five hundred million people every century in Europe alone. Its victims included sovereigns such as Czar Peter II of Russia, Queen Mary II of England, and Emperor Joseph I of Austria.

It was not until Columbus's arrival in the New World that Native Americans got their first exposure to the virus. The Europeans unwittingly brought a biological weapon with them that gave the invaders a brutal advantage over their opponents. With no immunity whatsoever to smallpox, Native Americans died in droves when they were exposed to the virus. In Central America, over 90 percent of the native population is believed to have died of smallpox in the decades following the arrival of the Spanish conquistadores in the early 1500s.

The first effective way to prevent the spread of smallpox probably arose in China around AD 900. A physician would rub a scab from a smallpox victim into a scratch in the skin of a healthy person. (Sometimes they administered it as an inhaled powder instead.) Variolation, as this process came to be called, typically caused just a single pustule to form on the inoculated arm. Once the pustule scabbed over, a variolated person became immune to smallpox.

At least, that was the idea. Fairly often, variolation would trigger more pustules, and in 2 percent of cases, people died. Still, a 2 percent risk was more attractive than the 30 percent risk of dying from a full-blown case of smallpox. Variolation spread across Asia, moving west along trade routes until the practice came to Constantinople in the 1600s. As news of its success traveled into Europe, physicians there began to practice variolation as well. The practice triggered religious objections that only God should decide who survived the dreaded smallpox. To counteract these suspicions, doctors organized public experiments. Zabdiel Boylston, a Boston doctor, publicly variolated hundreds of people in 1721 during a smallpox epidemic; those who had been variolated survived the epidemic in greater numbers than those who had not been part of the trial.

No one at the time knew why variolation worked, because nobody knew what viruses were or how our immune systems fought them. The treatment of smallpox moved forward mainly by trial and error. In the late 1700s, the British physician Edward Jenner invented a safer smallpox vaccine based on stories he heard about how milkmaids never got smallpox. Cows can get infected with cowpox, a close relative of smallpox, and so Jenner wondered if it provided some protection. He took pus from the hand of a milkmaid named Sarah Nelmes and inoculated it into the arm of a boy. The boy developed a few small pustules, but otherwise he suffered no symptoms. Six weeks later, Jenner variolated the boy—in other words, he exposed the boy to smallpox, rather than cowpox. The boy developed no pustules at all.

Jenner introduced the world to this new, safer way to prevent smallpox in a pamphlet he published in 1798. He dubbed it "vaccination," after the Latin name of cowpox, *Variolae vaccinae*. Within three years, over one hundred thousand people in England had gotten vaccinated against smallpox, and vaccinations spread around the world. In later years, other scientists borrowed Jenner's techniques and invented vaccines for other viruses. From rumors about milkmaids came a medical revolution.

As vaccines grew popular, doctors struggled to keep up with the demand. At first they would pick off the scabs that formed on vaccinated arms, and use them to vaccinate others in turn. But since cowpox occurred naturally only in Europe, people in other parts of the world could not simply acquire the virus themselves. In 1803, King Carlos of Spain came up with a radical solution: a vaccine expedition to the Americas and Asia. Twenty orphans boarded a ship in Spain. One of the orphans had been vaccinated before the ship set sail. After eight days, the orphan developed pustules, and then scabs. Those scabs were used to vaccinate another orphan, and so on through a chain of vaccination. As the ship stopped in port after port, the expedition delivered scabs to vaccinate the local population.

Physicians struggled throughout the 1800s to find a better way to deliver smallpox vaccines. Some turned calves into vaccine factories, infecting them repeatedly with cowpox. Some experimented with preserving the scabs in fluids like glycerol. It wasn't until scientists finally worked out the nature of smallpox—the fact that it was a virus—that it became possible to develop a vaccine that could be made on an industrial scale and shipped around the world.

Once vaccines became common, smallpox began to lose its fierce grip on humanity. Through the early 1900s, one country after another recorded their last case of smallpox. By 1959, smallpox had retreated from Europe, the Soviet Union, and North America. It remained a scourge of tropical countries with poor medical systems in place. But it was beaten so far back that some public health workers began to contemplate an audacious goal: eliminating smallpox from the planet altogether.

The advocates of smallpox eradication built their case on the biology of the virus. Smallpox infects only humans, not animals. If it could be systematically eliminated from every human population, there would be no need to worry that it was lurking in pigs or ducks, waiting to reinfect us. What's more, smallpox is an obvious disease. Unlike a virus like HIV, which can take years to make itself known, smallpox declares its gruesome presence in a matter of days. Public health workers would be able to identify outbreaks and track them with great precision.

Yet the idea of eradicating smallpox met with intense skepticism. If everything went exactly according to plan, an eradication project would require years of labor by thousands of trained workers, spread across much of the world, toiling in many remote, dangerous places. Public health workers had already tried to eradicate other diseases, like malaria, and failed.

The skeptics lost the debate, however, and in 1965, the World Health Organization launched the Intensified Smallpox Eradication Programme. The eradication effort was different in many ways from previous campaigns. It relied on a new prong-shaped needle that could deliver smallpox vaccine far more efficiently than regular syringes. As a result, vaccine supplies could be stretched much further than before. Public health workers also designed smart new strategies for administering vaccines. Trying to vaccinate entire countries was beyond the reach of the eradication project. Instead, public health workers identified outbreaks and took quick action to snuff them out. They quarantined victims and then vaccinated people in the surrounding villages and towns. The smallpox would spread like a forest fire, but soon it would hit the firebreak of vaccination and die out.

Outbreak by outbreak, the virus was beaten back, until the last case was recorded in Ethiopia in 1977. The world was now free of smallpox.

Ever since the campaign came to a close, it's served as proof that at least some pathogens can be wiped out. A few other campaigns have followed in its wake, but only one other virus has been eradicated successfully so far: the rinderpest virus. For centuries, rinderpest tormented dairy farmers and cattle-herders. Its fevers could kill off entire herds in one deadly sweep. Over the course of the 1900s, veterinarians carried out several campaigns against rinderpest, but they proved to be half-hearted, allowing the virus to bounce back again and again.

In the 1980s, rinderpest experts rethought their whole approach to the virus and began planning out a new campaign that would wipe it out for good. In 1990, vaccine developers created a cheap, stable rinderpest vaccine that could be transported on foot to even the most remote nomadic tribes. In 1994 the Food and Agricultural Organization used the vaccine to launch a global eradication program. They would gather information about sick cows from community workers and distribute vaccines where they were needed to keep infected animals from sickening healthy ones.

In country after country, rinderpest vanished. But wars would stop the campaigns and allow the virus to return to cleared territories. "Rinderpest is a prime candidate for eradication. Why has it not happened?" asked Sir Gordon Scott, a leader of the campaign, in a 1998 paper. "The major obstacle is 'man's inhumanity to man,'" he concluded. "Rinderpest thrives in a milieu of armed conflict and fleeing refugee masses."

Scott turned out to be far too pessimistic. In 2001, just three years after he wrote his gloomy predictions, veterinarians recorded their last case of rinderpest: a wild buffalo in Mount Meru National Park in Kenya. The FAO waited another decade to see whether any other animals fell ill. None did, and in 2011, they announced rinderpest had been eradicated.

Other eradication campaigns have come tantalizingly close to victory, only to bog down in the endgame. Polio, for example, was once a worldwide threat, leaving millions of children paralyzed or trapped in iron lungs. Years of eradication efforts have eliminated the virus from much of the world. In 1988, a thousand people developed new polio infections every day. In 2014, only one did each day. In 1988, polio was endemic in 125 countries. In 2014 it remained endemic in only three: Afghanistan, Nigeria, and Pakistan.

In all three countries, polio withstood eradication thanks to the disruptions of war and poverty. And in Pakistan, another threat developed: Taliban insurgents began to view vaccine campaigns as a threat and systematically assassinated vaccine workers. Pakistan had driven down its polio rates from 2,500 to 3,000 cases per year to a low of 28 in 2005. But then the rate started claiming again, reaching 93 in 2013 and then leaping to 327 in 2014. That year, the World Health Organization declared a polio public health emergency, as the virus turned up in outbreaks in countries that believed they were free of the virus—countries including Syria, Israel, Somalia, and Iraq.

As we begin eradicating viruses, we're also discovering that they can endure in all sorts of unnerving ways. In the late 1900s, as smallpox eradication workers traveled the world to wipe out the virus, scientists were also breeding it in their laboratories in order to study it. When the World Health Organization officially declared smallpox eradicated in 1980, those laboratory stocks remained. All it would take to reverse that eradication was for someone to accidentally set the virus loose.

The World Health Organization decided that all the laboratory stocks would eventually have to be destroyed. But in the interim, they would still allow scientists to conduct research on the virus, under strict World Health Organization rules. Scientists working with smallpox either had to destroy their stocks or send them to two approved laboratories, one in the Siberian city of Novosibirsk in the Soviet Union, and one at the US Centers for Disease Control and Prevention in Atlanta, Georgia. Over the next three decades, smallpox research continued under the World Health Organization's watchful eye. Scientists learned how to engineer lab animals to become infected with smallpox, allowing them to better understand its biology. They analyzed its genome, worked on better vaccines, and found drugs that showed promise as cures for smallpox. And during that time, the WHO debated exactly when they should destroy the virus once and for all.

Some experts argued that there was no reason to wait. As long as smallpox existed—no matter how carefully controlled—the risk remained that the virus could escape and kill millions of people. Terrorists might even try to use it as a biological weapon. Making matters worse, people were no longer getting smallpox vaccines, and so what immunity the public had to the virus was waning.

But other scientists urged holding onto smallpox stocks. They pointed out that the eradication campaign might not, in fact, have been a complete success. In the 1990s, Soviet defectors revealed that their government had set up labs to produce a weaponized form of smallpox, one that could be loaded in missiles and launched at enemy targets. After the fall of the Soviet Union, those biological warfare labs were abandoned. No one knows what ultimately happened to the smallpox viruses used for that research. We are left with the terrifying possibility that ex-Soviet virologists sold smallpox stocks to other governments or even terrorist organizations.

In 2014, the world discovered that you don't need a biological warfare program to lose track of smallpox. Scientists at the National Institutes of Health in Maryland were packing up the contents of a lab when they came across half a dozen old vials. The vials, which dated back from the 1950s, contained smallpox. In the World Health Organization's sweep of smallpox, those vials had gone overlooked, despite being stored in one of the world's leading medical research centers.

Opponents of smallpox eradication argue that the risk of new outbreaks—no matter how small—justifies more research on the virus. There's still so much we don't yet know about it. Smallpox can infect only a single species—humans—while all its relatives, called orthopoxes, can infect several species. No one knows what makes smallpox special. If a smallpox outbreak should occur in years to come, fast diagnosis could save untold lives. But the current tests for smallpox are based on obsolete technology. In order to develop cutting-edge diagnostics, scientists will need to test them to be sure that they distinguish between smallpox and other orthopox viruses. Only live smallpox viruses will do for such experiments. Likewise, scientists could use the viruses to develop better vaccines and antiviral drugs.

The debate over smallpox did not resolve itself in a clean decision—only an agreement to bring up the matter in the future. But as the disagreement dragged on, advances in technology changed the very terms of the debate.

In the 1970s, scientists developed the first methods for sequencing the genetic material in organisms. DNA is a pair of strands assembled from units known as bases, while RNA is a single strand. There are four different kinds of bases in genes, acting something like a four-letter alphabet. All of the DNA in a human cell—the human genome—would come to 3.2 billion base pairs. If you printed a letter for each base in the genome, it would fill a book a thousand times longer than *War and Peace*. Because the initial methods for sequencing DNA were so slow and unreliable, scientists started sequencing genomes with the smallest genomes on Earth: those of viruses. In 1976, they published the genome of a bacteriophage called MS2—a mere 3,569 bases.

In the years that followed, scientists published the genomes of other viruses, including smallpox in 1993. By comparing its genome to those of other viruses, the scientists gained some clues to the workings of its proteins. Researchers went on to sequence the genomes of smallpox strains from other parts of the world, revealing that there was little variation between them—an important clue for researchers planning on preparations for future smallpox outbreaks.

The invention of genome-sequencing technology opened the way for another major advance: scientists began to assemble bases to synthesize genes from scratch. At first they assembled short stretches of genetic material. Even at that early stage, Eckerd Wimmer, a virologist at Stony Brook University, realized that viruses had genomes that were small enough that they could be synthesized in full. In 2002, he and his colleagues used the poliovirus genome as a guide for the creation of thousands of short DNA fragments. They then used enzymes to stitch the fragments together, and used the DNA molecule as a template for making a corresponding RNA molecule—in other words, a physical copy of the entire poliovirus genome. When Wimmer and his colleagues added that RNA to test tubes full of bases and enzymes, live polioviruses spontaneously assembled. They had, in other words, made polio from scratch.

Smallpox is a large virus—about 25 times bigger than polio—

and its enormous size would create a big challenge for any scientist who wanted to synthesize it from scratch. But it's conceivable that a large, well-trained lab could pull it off, given enough time. There's no evidence that anyone has tried to resurrect smallpox using Wimmer's approach, but, then again, there's no evidence that it would be impossible to do so. After thirty-five hundred years of suffering and puzzling over smallpox, we have finally come to understand it and halt its destruction. And yet, by understanding smallpox, we have ensured that it can never be utterly eradicated as a threat to humans. Our growing knowledge of viruses has given smallpox its own kind of immortality.

EPILOGUE

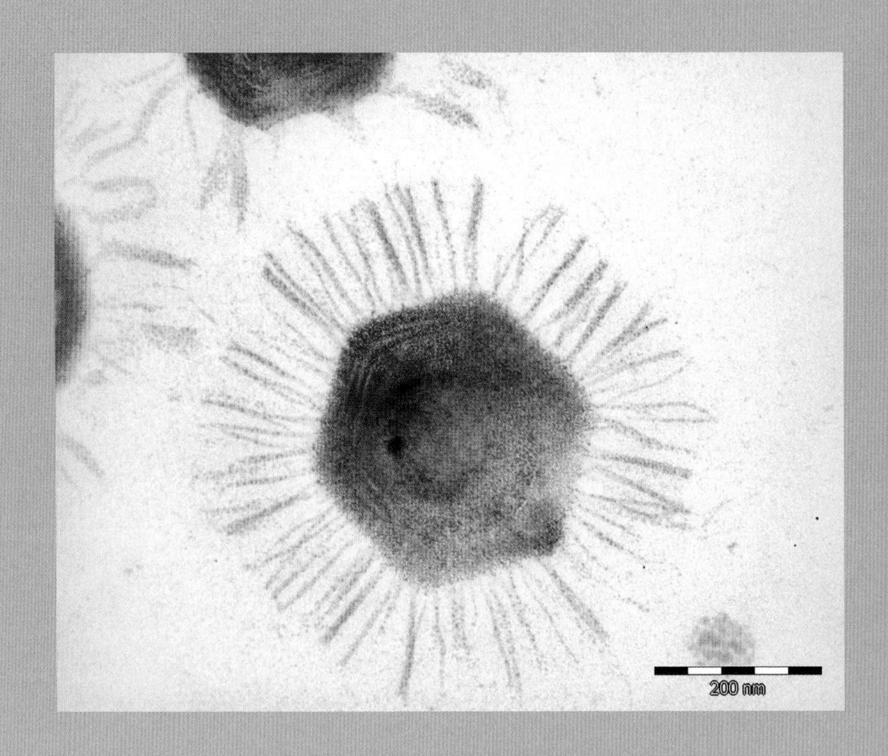

Mimivirus, one of the largest known viruses

The Alien in the Water Cooler

Giant Viruses and the Definition of Life

Wherever there is water on Earth, there is life. The water may be a Yellowstone geyser, a pool in the Cave of Crystals, or a cooling tower sitting on the roof of a hospital.

In 1992, a microbiologist named Timothy Rowbotham scooped up some water from a hospital cooling tower in the English city of Bradford. He put it under a microscope and saw a welter of life. He saw amoebae and other single-celled protozoans, about the size of human cells. He saw bacteria, about a hundred times smaller. Rowbotham was searching for the cause of an outbreak of pneumonia that had been raging through Bradford. In the ranks of the microbes he found in the cooling tower water, he thought he found a promising candidate: a sphere of bacterial size, sitting inside an amoeba. Rowbotham believed he had found a new bacterium, and dubbed it *Bradfordcoccus*.

Rowbotham spent years trying to make sense of *Bradfordcoccus*, to see whether it was the culprit in the pneumonia outbreak. He tried to identify its genes by finding matches with genes in other species of bacteria. But he couldn't find any. Rowbotham had reached a dead end by 1998, when budget cuts forced him to close his lab down. Not wanting to destroy his puzzling *Bradfordcoccus*, he arranged for French colleagues to store his samples.

For years, *Bradfordcoccus* languished in obscurity, until Bernard La Scola of Mediterranean University decided to take another look at it. As soon as he put Rowbotham's samples under a microscope, he realized something was not right.

Bradfordcoccus did not have the smooth surface of spherical bacteria. Instead, it was more like a soccer ball, made up of many interlocking plates. And La Scola saw hairlike threads of protein radiating out from its geometric shell. The only things in nature known to have these kinds of shells and threads were viruses. But La Scola knew, like all microbiologists at the time knew, that something the size of *Bradfordcoccus* was a hundred times too big to be a virus.

And yet a virus is exactly what *Bradfordcoccus* turned out to be. When La Scola and his colleagues examined it further, they discovered that it reproduced by invading amoebae and forcing them to build new copies of itself. Only viruses reproduce this way. La Scola and his colleagues gave *Bradfordcoccus* a new name to reflect its viral nature. They called it a mimivirus, in honor of the virus's ability to mimic bacteria.

The French scientists set out to analyze the genes of the mimivirus. Rowbotham had tried—and failed—to match its genes to those of bacteria. The French scientists had better luck. The mimivirus turned out to have virus genes—and a lot of them. Before the discovery of mimiviruses, scientists had become accustomed to finding only a few genes in a virus. But mimiviruses have 1,018 genes. It was as if someone took the genomes of the flu, the cold, smallpox, and a hundred other viruses, and stuffed them all inside one protein shell. The mimivirus even had more genes than some species of bacteria. In both its size and its genes, mimivirus had broken cardinal rules for being a virus.

La Scola and his colleagues published the details of their remarkable mimivirus in 2003. They wondered whether it was unique or whether there were other giant viruses also hidden in plain sight. They collected water from cooling towers in France and added amoebae to it, to see whether anything in the water might infect them. Soon the amoebae were exploding, releasing giant viruses. These were not mimiviruses, however. They were another species, with 1,059 genes, setting a new record for the biggest genome in a virus. While the new virus looked a lot like mimiviruses, its genome was profoundly different. When the scientists lined up the new virus's genes with those of mimivirus, they could find matches for 833 of them. But the other 226 were unique. The scientists decided their new giant virus needed a name. They called it mamavirus.

Other researchers joined in on the hunt for giant viruses, and they began finding them everywhere: in rivers, in oceans, in lakes buried under Antarctic ice. In 2014, French researchers thawed out pieces of Siberian permafrost that had been frozen for 30,000 years. They discovered giant viruses in the thawed soil—at 1.5 micrometers long, they're the biggest giant viruses yet found.

Scientists have even begun finding giant viruses lurking in animals. La Scola and his colleagues collaborated with Brazilian scientists to study serum samples taken from mammals. They found antibodies to giant viruses in monkeys and cows. The researchers have also isolated giant viruses themselves from people, including a patient with pneumonia. It's not clear yet what role giant viruses play in our health, though. They may be able to directly infect our own cells, or they may just lurk harmlessly in amoebae that invade our bodies.

The story of giant viruses drives home just how little of the virosphere we have explored. It also injects fresh life into a long-standing debate that began over a century ago with the discovery of viruses: what does it mean to be alive?

Before scientists discovered viruses, they generally agreed on the definition of life. Living things could use their metabolism to survive, grow, and reproduce. Other states of matter, like clouds or crystals, might seem alive in one respect or another, but on the whole they didn't come close to meeting all the standards for life.

When Wendell Stanley produced crystals of tobacco mosaic virus in 1935, he unsettled those neat distinctions between the living and the nonliving. As a crystal, his virus behaved like ice or diamonds. But when provided with a tobacco plant, it multiplied like any living thing.

Later, as the molecular biology of viruses came into crisper focus, many scientists decided that they were only lifelike, but not truly alive. All the viruses that scientists studied carried a few genes apiece, leaving a wide genetic gulf dividing them from bacteria. The few genes that viruses carried allowed them to execute the barest tasks required for making new viruses: to invade a cell and slip their genes into a cell's biochemical factories. Missing from viruses were all the genes for full-blown life. Scientists could find no instructions in a virus for making a ribosome, for example, the molecular factory that turns RNA into proteins. Nor did viruses have genes for the enzymes that break down food in order to grow. In other words, viruses appeared to lack much of the genetic information required to be truly alive.

In theory, though, a virus might be able to gain that information and become truly alive. After all, viruses are not fixed in stone. A mutation might accidentally duplicate some of their genes, creating new copies that could take on new functions. Or a virus might accidentally take up genes from another virus, or even from a host cell. Its genome could expand until it could feed, grow, and divide on its own.

While it was conceivable that viruses could evolve their way to life, scientists saw a giant wall in their way. Organisms with big genomes need a way to copy them accurately. The odds of suffering a dangerous mutation increase as a genome gets larger. We protect our giant genomes from this risk by producing error-correcting enzymes, as do other animals, plants, fungi, protozoans, and bacteria. Viruses, on the other hand, have no repair enzymes. As a result, they make copying errors at a tremendously higher rate than we do—in some cases, over a thousand times higher.

The high mutation rate of viruses may put a limit on their genome—and thus may prevent them from being truly alive. If a virus's genome gets bigger, it is more likely to suffer a lethal mutation. Natural selection may therefore favor tiny genomes in viruses. If that's true, then viruses may be unable to make room for genes that would let them turn raw ingredients into new genes and proteins. They cannot grow. They cannot expel waste. They cannot defend against hot and cold. They cannot reproduce by splitting in two.

All those *nots* added up to one great, devastating *NOT*. Viruses were not alive.

"An organism is constituted of cells," the microbiologist Andre Lwoff declared in a lecture he gave when he accepted the Nobel Prize in 1967. Not being cells, viruses were considered little more than cast-off genetic material that happened to have the right chemistry to get replicated inside cells. In 2000, the International Committee on Taxonomy of Viruses made this judgment official. "Viruses are not living organisms," they flatly declared.

But in the years following that declaration, many virologists came to question it, and some rejected it outright. In the face of new discoveries, the old rules no longer work well. Giant viruses, for example, went overlooked for so long in part because they were a hundred times bigger than viruses are supposed to be. They are also loaded with far too many genes to fit old-fashioned notions of a virus. Scientists don't know what giant viruses do with all of their genes, but some suspect that they do some rather lifelike things with them. Some genes in giant viruses encode enzymes that can repair DNA. They may use these enzymes to fix any damage they incur while traveling from one host cell to another. And when giant viruses invade amoebae, they don't dissolve into a cloud of molecules. Instead, they set up a massive, intricate structure called a viral factory. The viral factory takes in raw ingredients through one portal, and then spits out new DNA and proteins through two others. Giant viruses may use their viral genes to carry out at least some of this biochemical work.

The giant virus's viral factory, in other words, looks and acts remarkably like a cell. It's so much like a cell, in fact, that La Scola and his colleagues discovered in 2008 that it can be infected by a virus of its own. This new kind of virus, which they named a virophage, slips into the virus factory and fools it into building virophages instead of giant viruses.

Drawing dividing lines through nature can be scientifically useful, but when it comes to understanding life itself, those lines can end up being artificial barriers. Rather than trying to figure out how viruses are not like other living things, it may be more useful to think about how viruses and other organisms form a continuum. We humans are an inextricable blend of mammal and virus. Remove our virus-derived genes, and we would die in the womb. It's also likely that we depend on our viral DNA to defend against infections. Some of the oxygen we breathe is produced through a mingling of viruses and bacteria in the oceans. That mixture is not a fixed combination, but an ever-changing flux. The oceans are a living matrix of genes, shuttling among hosts and viruses.

While it's clear that giant viruses bridge the gap between ordinary viruses and cellular life, it's not yet clear how they reached that ambiguous position. Some researchers argue that they started out as ordinary viruses and stole their extra genes from their hosts. Others have countered that giant viruses started out early in the history of life as free-living cells, and became more virus-like over billions of years.

Drawing a bright line between life and nonlife doesn't just make it harder to understand viruses. It also makes it harder to appreciate how life began. Scientists are still trying to work out the origin of life, but one thing is clear: it did not start suddenly with the flick of a great cosmic power switch. It's likely that life emerged gradually, as raw ingredients like sugar and phosphate combined in increasingly complex reactions on the early Earth. It's possible, for example, that single-stranded molecules of RNA gradually grew and acquired the ability to make copies of themselves. Trying to find a moment in time when such RNA-life abruptly became "alive" just distracts us from the gradual transition to life as we know it.

In the RNA world, life may have consisted of little more than fleeting coalitions of genes, which sometimes thrived and sometimes were undermined by genes that acted like parasites. Some of those primordial parasites may have evolved into the first viruses, which may have continued replicating down to the present day. Patrick Forterre, a French virologist, has proposed that in the RNA world, viruses invented the double-stranded DNA molecule as a way to protect their genes from attack. Eventually their hosts took over their DNA, which then took over the world. Life as we know it, in other words, may have needed viruses to get its start.

At long last, we may be returning to the original two-sided sense of the word virus, which originally signified either a life-giving substance or a deadly venom. Viruses are indeed exquisitely deadly, but they have provided the world with some of its most important innovations. Creation and destruction join together once more.

Acknowledgments

A Planet of Viruses was funded by the National Center for Research Resources at the National Institutes of Health through the Science Education Partnership Award (SEPA), grant no. R25 RR024267 (2007–2012), Judy Diamond, Moira Rankin, and Charles Wood, principal investigators. Its content is solely the responsibility of the author and does not necessarily represent the official views of the NCRR or the NIH. I thank the many people who advised this project: Anisa Angeletti, Peter Angeletti, Aaron Brault, Ruben Donis, Ann Downer-Hazell, David Dunigan, Angie Fox, Laurie Garrett, Benjamin David Jee, Ian Lipkin, Ian Mackay, Grant McFadden, Nathan Meier, Abbie Smith, Gavin Smith, Philip W. Smith, Amy Spiegel, David Uttal, James L. Van Etten, Kristin Watkins, Willie Wilson, and Nathan Wolfe. I am particularly grateful to my SEPA program officer, L. Tony Beck, and to my editor at the University of Chicago Press, Christie Henry, for making this book possible.

Selected References

A CONTAGIOUS LIVING FLUID

Bos, L. 1999. Beijerinck' s work on tobacco mosaic virus: Historical context and legacy. *Philosophical Transactions of the Royal Society B: Biological Sciences* 354:675.

Flint, S. J. 2009. *Principles of Virology*. 3rd ed. Washington, DC: ASM Press.

Kay, L. E. 1986. W. M. Stanley's crystallization of the tobacco mosaic virus, 1930–1940. *Isis* 77:450–72.

Willner D., M. Furlan, M. Haynes, et al. 2009. Metagenomic analysis of respiratory tract DNA viral communities in cystic fibrosis and non-cystic fibrosis individuals. *PLoS ONE* 4 (10):e7370.

THE UNCOMMON COLD

Arden, K. E., and I. M. Mackay. 2009. Human rhinoviruses: Coming in from the cold. *Genome Medicine* 1:44.

Briese, T., N. Renwick, M. Venter, et al. 2008. Global distribution of novel rhinovirus genotype. *Emerging Infectious Diseases* 14:944.

Fashner, J., K. Ericson, and S. Werner. 2012. Treatment of the common cold in children and adults. *American Family Physician* 86:153–59.

Palmenberg, A. C., D. Spiro, R. Kuzmickas, et al. 2009. Sequencing and analyses of all known human rhinovirus genomes reveal structure and evolution. *Science* 324:55–59.

LOOKING DOWN FROM THE STARS

Barry, J. M. 2004. *The Great Influenza: The Epic Story of the Deadliest Plague in History*. New York: Viking.

Dugan, V. G., R. Chen, D. J. Spiro, et al. 2008. The evolutionary genetics and emergence of avian influenza viruses in wild birds. *PLoS Pathogens* 4 (5):e1000076.

Rambaut, A., O. G. Pybus, M. I. Nelson, C. Viboud, J. K. Taubenberger, and E. C. Holmes. 2008. The genomic and epidemiological dynamics of human influenza A virus. *Nature* 453:615–19.

Smith, G. J. D., D. Vijaykrishna, J. Bahl, et al. 2009. Origins and evolutionary genomics of the 2009 swine-origin H1N1 influenza A epidemic. *Nature* 459:1122–25.

Taubenberger, J. K., and D. M. Morens. 2008. The pathology of influenza virus infections. *Annual Reviews of Pathology* 3:499–522.

RABBITS WITH HORNS

Bravo, I. G., and Á. Alonso. 2006. Phylogeny and evolution of papillomaviruses based on the E1 and E2 proteins. *Virus Genes* 34:249–62.

Doorbar, J. 2006. Molecular biology of human papillomavirus infection and cervical cancer. *Clinical Science* 110:525.

García-Vallvé, S., Á. Alonso, and I. G. Bravo. 2005. Papillomaviruses: Different genes have different histories. *Trends in Microbiology* 13:514–21.

García-Vallvé, S., J. R. Iglesias-Rozas, Á. Alonso, and I. G. Bravo. 2006. Different papillomaviruses have different repertoires of transcription factor binding sites: Convergence and divergence in the upstream regulatory region. *BMC Evolutionary Biology* 6:20.

Horvath, C. A. J., G. A. V. Boulet, V. M. Renoux, P. O. Delvenne, and J.-P. J. Bogers. 2010. Mechanisms of cell entry by human papillomaviruses: An overview. *Virology Journal* 7:11.

Martin, D., and J. S. Gutkind. 2008. Human tumor-associated viruses and new insights into the molecular mechanisms of cancer. *Oncogene* 27 (Suppl 2):S31–42.

Orlando, P. A., R. A. Gatenby, A. R. Giuliano, and J. S. Brown. 2012. Evolutionary ecology of human papillomavirus: trade-offs, coexistence, and origins of high-risk and low-risk types. *Journal of Infectious Diseases* 205:272–79.

Schiffman, M., R. Herrero, R. DeSalle, et al. 2005. The carcinogenicity of human papillomavirus types reflects viral evolution. *Virology* 337:76–84.

Shulzhenko, N., H. Lyng, G. F. Sanson, and A. Morgun. 2014. Ménage à trois: An evolutionary interplay between human papillomavirus, a tumor, and a woman. *Trends in Microbiology* 22:345–53.

THE ENEMY OF OUR ENEMY

Adhya, S., C. R. Merril, and B. Biswas. 2014. Therapeutic and prophylactic applications of bacteriophage components in modern medicine. *Cold Spring*

Harbor Perspectives in Medicine 4:a012518.
Brussow, H. 2005. Phage therapy: The *Escherichia coli* experience. *Microbiology* 151:2133.
Deresinski, S. 2009. Bacteriophage therapy: Exploiting smaller fleas. *Clinical Infectious Diseases* 48:1096–101.
Summers, W. C. 2001. Bacteriophage therapy. *Annual Reviews in Microbiology* 55:437–51.

THE INFECTED OCEAN

Angly, F. E., B. Felts, M. Breitbart, et al. 2006. The marine viromes of four oceanic regions. *PLoS Biology* 4 (11):e368.
Breitbart, M. 2012. Marine viruses: Truth or dare. *Annual Review of Marine Sciences* 4:425–48.
Brussaard, C. P. D., S. W. Wilhelm, F. Thingstad, et al. 2008. Global-scale processes with a nanoscale drive: The role of marine viruses. *ISME Journal* 2:575–78.
Danovaro, R., A. Dell'Anno, C. Corinaldesi, et al. 2008. Major viral impact on the functioning of benthic deep-sea ecosystems. *Nature* 454:1084–87.
Desnues, C., B. Rodriguez-Brito, S. Rayhawk, et al. 2008. Biodiversity and biogeography of phages in modern stromatolites and thrombolites. *Nature* 452:340–43.
Keen, E.C. 2015. A century of phage research: Bacteriophages and the shaping of modern biology. *Bioessays* 37:6–9.
Rohwer, F., and R. Vega Thurber. 2009. Viruses manipulate the marine environment. *Nature* 459:207–12.
Rohwer, F., M. Youle, H. Maughan, and N. Hisakawa. 2015. *Life in Our Phage World*. San Diego: Wholon.
Suttle, C. A. 2007. Marine viruses—major players in the global ecosystem. *Nature Reviews Microbiology* 5:801–12.
Van Etten, J. L., L. C. Lane, and R. H. Meints. 1991. Viruses and viruslike particles of eukaryotic algae. *Microbiology and Molecular Biology Reviews* 55:586.

OUR INNER PARASITES

Blikstad, V., F. Benachenhou, G. O. Sperber, and J. Blomberg. 2008. Evolution of human endogenous retroviral sequences: A conceptual account. *Cellular and Molecular Life Sciences* 65:3348–65.
Dewannieux, M., F. Harper, A. Richaud, et al. 2006. Identification of an infectious progenitor for the multiple-copy HERV-K human endogenous retroelements. *Genome Research* 16:1548–56.
Jern, P., and J. M. Coffin. 2008. Effects of retroviruses on host genome function. *Annual Review of Genetics* 42:709–32.
Lavialle, C., G. Cornelis, A. Dupressoir, et al. 2013. Paleovirology of "syncytins," retroviral env genes exapted for a role in placentation. *Philosophical Transactions of the Royal Society B: Biological Sciences* 368:20120507.
Lee, A., A. Nolan, J. Watson, and M. Tristem. 2013. Identification of an ancient

endogenous retrovirus, predating the divergence of the placental mammals. *Philosophical Transactions of the Royal Society B: Biological Sciences* 368:20120503.

Ruprecht, K., J. Mayer, M. Sauter, K. Roemer, and N. Mueller-Lantzsch. 2008. Endogenous retroviruses and cancer. *Cellular and Molecular Life Sciences* 65:3366–82.

Tarlinton, R., J. Meers, and P. Young. 2008. Biology and evolution of the endogenous koala retrovirus. *Cellular and Molecular Life Sciences* 65:3413–21.

Weiss, R. A. 2006. The discovery of endogenous retroviruses. *Retrovirology* 3:67.

Weiss, R. A. 2013. On the concept and elucidation of endogenous retroviruses. *Philosophical Transactions of the Royal Society B: Biological Sciences* 368:20120494.

THE YOUNG SCOURGE

D'arca, M., A. Ayoubaa, A. Estebana, et al. 2015. Origin of the HIV-1 group O epidemic in western lowland gorillas. *Proceedings of the National Academy of Sciences* Published ahead of print, March 2, 2015. doi:10.1073/pnas.1502022112.

Fan, H. 2011. *AIDS: Science and Society*. 6th ed. Sudbury, MA: Jones and Bartlett.

Faria, N. R., A. Rambaut, M. A. Suchard, et al. 2014. The early spread and epidemic ignition of HIV-1 in human populations. *Science* 346:56–61.

Fauci, A. S., and H. D. Marston. 2014. Ending AIDS–is an HIV vaccine necessary? *New England Journal of Medicine* 370:495–98.

Gilbert, M. T. P., A. Rambaut, G. Wlasiuk, T. J. Spira, A. E. Pitchenik, and M. Worobey. 2007. The emergence of HIV/AIDS in the Americas and beyond. *Proceedings of the National Academy of Sciences* 104:18566.

Keele, B. F. 2006. Chimpanzee reservoirs of pandemic and nonpandemic HIV-1. *Science* 313:523–26.

Montagnier, L. 2010. 25 Years after HIV discovery: Prospects for cure and vaccine. *Virology* 397:248–54.

Worobey, M., M. Gemmel, D. E. Teuwen, et al. 2008. Direct evidence of extensive diversity of HIV-1 in Kinshasa by 1960. *Nature* 455:661–64.

BECOMING AN AMERICAN

Brault, A. C. 2009. Changing patterns of West Nile virus transmission: Altered vector competence and host susceptibility. *Veterinary Research* 40:43.

Diamond, M. S. 2009. Progress on the development of therapeutics against West Nile virus. *Antiviral Research* 83:214–27.

Gould, E. A., and S. Higgs. 2009. Impact of climate change and other factors on emerging arbovirus diseases. *Transactions of the Royal Society of Tropical Medicine and Hygiene* 103:109–21.

Hamer, G. L, U. D. Kitron, T. L. Goldberg, et al. 2009. Host selection by *Culex pipiens* mosquitoes and West Nile virus amplification. *American Journal of Tropical Medicine and Hygiene* 80:268.

Sfakianos, J. N. 2009. *West Nile Virus*. 2nd ed. New York: Chelsea House.
Venkatesan M., and J. L. Rasgon. 2010. Population genetic data suggest a role for mosquito-mediated dispersal of West Nile virus across the western United States. *Molecular Ecology* 19:1573–84.

PREDICTING THE NEXT PLAGUE

Holmes, E. C., and A. Rambaut. 2004. Viral evolution and the emergence of SARS coronavirus. *Philosophical Transactions of the Royal Society B: Biological Sciences* 359:1059–65.
Parrish, C. R., E. C. Holmes, D. M. Morens, et al. 2008. Cross-species virus transmission and the emergence of new epidemic diseases. *Microbiology and Molecular Biology Reviews* 72:457–70.
Pigott, D. M., N. Golding, A. Mylne, et al. 2014. Mapping the zoonotic niche of Ebola virus disease in Africa. *Elife* 3:e04395.
Piot, P. 2013. *No Time to Lose: A Life in Pursuit of Deadly Viruses*. New York: W. W. Norton.
Quammen, D. 2012. *Spillover: Animal Infections and the Next Human Pandemic*. New York: W. W. Norton.
Raj, V. S., A. D. Osterhaus, R. A. Fouchier, and B. L. Haagmans. 2014. MERS: Emergence of a novel human coronavirus. *Current Opinion in Virology* 5:58–62.
Sack, K., S. Fink, P. Belluck, and A. Nossiter. 2014. Ebola's deadline escape. *New York Times*, December 30, 2014, D1.
Skowronski, D. M., C. Astell, R. C. Brunham, et al. 2005. Severe acute respiratory syndrome (SARS): A year in review. *Annual Review of Medicine* 56:357–81.
WHO Ebola Response Team. 2014. Ebola virus disease in West Africa—the first 9 months of the epidemic and forward projections. *New England Journal of Medicine* 371:1481–95.
Wolfe, N. 2009. Preventing the next pandemic. *Scientific American*, April 2009, 76–81.

THE LONG GOODBYE

Damon, I. K., C. R. Damaso, and G. McFadden. 2014. Are we there yet? The smallpox research agenda using variola virus. *PLoS Pathogens* 10:e1004108.
Dormitzer, P. R., P. Suphaphiphat, D. G. Gibson, et al. 2013. Synthetic generation of influenza vaccine viruses for rapid response to pandemics. *Science Translational Medicine* 5:185ra68.
Hughes, A. L., S. Irausquin, and R. Friedman. 2010. The evolutionary biology of poxviruses. *Infection, Genetics and Evolution* 10:50–59.
Jacobs, B. L., J. O. Langland, K. V. Kibler, et al. 2009. Vaccinia virus vaccines: Past, present and future. *Antiviral Research* 84:1–13.
Kennedy, R. B., I. Ovsyannikova, and G. A. Poland. 2009. Smallpox vaccines for biodefense. *Vaccine* 27 (Suppl):D73–79.
Koplow, D. A. 2003. *Smallpox: The Fight to Eradicate a Global Scourge*. Berkeley: University of California Press.

Kosuri, S., and G. M. Church. 2014. Large-scale de novo DNA synthesis: Technologies and applications. *Nature Methods* 11:499–507.

Mariner, J. C., J. A. House, C. A. Mebus, et al. 2012. Rinderpest eradication: Appropriate technology and social innovations. *Science* 337:1309–12.

Reardon, S. 2014. "Forgotten" NIH smallpox virus languishes on death row. *Nature* 514:544.

Shchelkunov, S. N. 2009. How long ago did smallpox virus emerge? *Archives of Virology* 154:1865–71.

Wimmer, E. 2006. The test-tube synthesis of a chemical called poliovirus. *EMBO Reports* 7:S3–9.

THE ALIEN IN THE WATER COOLER

Abrahão, J. S., F. P. Dornas, L. C. F. Silva, et al 2014. *Acanthamoeba polyphaga mimivirus* and other giant viruses: an open field to outstanding discoveries. *Virology Journal* 11:120.

Claverie, J. M., and C. Abergel. 2013. Open questions about giant viruses. *Advances in Virus Research* 85:25-56.

Forterre, P. 2010. Defining life: The virus viewpoint. *Origins of Life and Evolution of Biospheres* 40:151–60.

Forterre, P., M. Krupovic, and D. Prangishvili. 2014. Cellular domains and viral lineages. *Trends in Microbiology* 22:554–58.

Katzourakis, A., and A. Aswad. 2014. The origins of giant viruses, virophages and their relatives in host genomes. *BMC Biology* 12:51.

Koonin, E. V., and V. V. Dolja. 2014. Virus world as an evolutionary network of viruses and capsidless selfish elements. *Microbiology and Molecular Biology Reviews* 78:278–303.

Moreira, D., and C. Brochier-Armanet. 2008. Giant viruses, giant chimeras: The multiple evolutionary histories of mimivirus genes. *BMC Evolutionary Biology* 8:12.

Moreira, D., and P. Lopez-Garcia. 2009. Ten reasons to exclude viruses from the tree of life. *Nature Reviews Microbiology* 7:306–11.

Ogata, H., and J. M. Claverie. 2008. How to infect a mimivirus. *Science* 321:1305.

Raoult, D., and P. Forterre. 2008. Redefining viruses: Lessons from mimivirus. *Nature Reviews Microbiology* 6:315–19.

Raoult, D., B. La Scola, and R. Birtles. 2007. The discovery and characterization of mimivirus, the largest known virus and putative pneumonia agent. *Clinical Infectious Diseases* 45:95–102.

Credits

Introduction: tobacco mosaic viruses, © Dennis Kunkel Microscopy, Inc. Chapter 1: rhinovirus, copyright © 2010 Photo Researchers, Inc. (all rights reserved). Chapter 2: influenza virus, by Frederick Murphy, from the PHIL, courtesy of the CDC. Chapter 3: human papillomavirus, copyright © 2010 Photo Researchers, Inc. (all rights reserved). Chapter 4: bacteriophages, courtesy of Graham Colm. Chapter 5: marine phage, courtesy of Willie Wilson. Chapter 6: avian leukosis virus, courtesy of Dr. Venugopal Nair and Dr. Pippa Hawes, Bioimaging group, Institute for Animal Health. Chapter 7: human immunodeficiency virus, by P. Goldsmith, E. L. Feorino, E. L. Palmer, and W. R. McManus, from the PHIL, courtesy of the CDC. Chapter 8: West Nile virus, by P. E. Rollin, from the PHIL, courtesy of the CDC. Chapter 9: Ebola virus, by Cynthia Goldsmith, from the PHIL, courtesy of the CDC. Chapter 10: smallpox virus, by Frederick Murphy, from the PHIL, courtesy of the CDC. Epilogue: mimivirus, courtesy of Dr. Didier Raoult, Research Unit in Infectious and Tropical Emergent Diseases (URMITE).

Index